Participatory Forestry: The Process of Change in India and Nepal

Mary Hobley

Participatory Forestry: The Process of Change in India and Nepal

Mary Hobley

Rural Development Forestry
Study Guide 3
Rural Development Forestry Network
Overseas Development Institute
London
1996

First published 1996

Published by the Overseas Development Institute,
Regent's College, Inner Circle, Regent's Park, London NW1 4NS

ISBN 0 85003 204 0

Typeset by Discript, London
Printed by Russell Press Ltd, Nottingham

Cataloguing in Publication Data for this book is available from the British Library.

The research for and production of this book has been supported and financed by:

The European Commission DG VIII Project No: Ref B7–5040/91/046
Forestry Research Programme, University of Oxford, Halifax House,
6 South Parks Road, Oxford OX1 3UB
Voluntary Service Overseas, 317 Putney Bridge Road, London
SW15 2PN, UK

Contents

List of Tables

List of Figures

List of Boxes

List of Photographs and Illustrations

All photographs taken by Mary Hobley

Acronyms

ACF	Assistant Conservator of Forests
AKRSP	Aga Khan Rural Support Programme
CCF	Chief Conservator of Forests
CF	Conservator of Forests
CPFD	Community and Private Forestry Department
CPR	Common Property Resources and Common Pool Resources
CSE	Centre for Science and Environment
DFO	District Forest Office(r)
DOF	Department of Forestry
FAO	Food and Agriculture Organisation
FECOFUN	Federation of Community Forestry Users of Nepal
FD	Forest Department
FG	Forest Guard
FLC	Forest Labour Cooperative
FPC	Forest Protection Committee
FRI	Forestry Research Institute
FSO	Forest Settlement Officer
FTPP	Forests, Trees and People Programme
FUG	Forest User Group
GIS	Geographical Information Systems
GO	Government Organisation
GOI	Government of India
GON	Government of Nepal
GVM	Gram Vikas Mandal
HMGN	His Majesty's Government of Nepal
HRMS	Hill Resource Management Society
IBRAD	Institute for Bio-Social Research and Development
ICIMOD	International Centre for Integrated Mountain Development
ICRAF	International Centre for Agroforestry Research
IFS	Indian Forest Service

IIFM	Indian Institute of Forest Management
IOF	Institute of Forestry
ITTO	International Tropical Timber Organisation
IUCN	International Union for the Conservation of Nature
JFM	Joint Forest Management
MOF	Ministry of Forests
MPFS	Master Plan for the Forest Sector
MSC	Multiple Shoot Cutting
N-AFP	Nepal-Australia Forestry Project
NACFP	Nepal-Australia Community Forestry Project
NCA	National Commission on Agriculture
NGO	Non Government Organisation
NTFP	Non Timber Forest Products
NUKCFP	Nepal-UK Community Forestry Project
NWDB	National Wastelands Development Board
NWFP	Non Wood Forest Products
ODA	Overseas Development Administration
ODI	Overseas Development Institute
PCCF	Principal Chief Conservator of Forests
PRA	Participatory Rural Appraisal
RDF	Rural Development Forestry
RFO	Range Forest Officer
RRA	Rapid Rural Appraisal
SIDA	Swedish International Development Administration
SIFPG	Self-initiated Forest Protection Group
SPWD	Society for the Promotion of Wastelands Development
TCN	Timber Corporation of Nepal
TERI	Tata Energy Research Institute
UMN	United Mission to Nepal
VDC	Village Development Committee
VFC	Village Forest Committee
VFPC	Village Forest Protection Committee
VIKSAT	Vikram Sarabhai Centre for Science and Technology
VLW	Village Level Worker
VSO	Voluntary Service Overseas

Foreword

Participatory forest management in India and Nepal is rooted in the history of people's movements for social and economic justice. This struggle shows three distinct historical trends. The first trend documented in this Study Guide has been the systematic alienation of forest communities from their resource by the state. The environmental consequences of the often indiscriminate exploitation of forests by the colonial state and ruling classes gave birth to the second major trend, of natural resource protection against the majority of local resource users. The third and most recent trend represents an accommodation between the interests of the state and people's movements for control over forests. Thus participatory forestry is an expression of the contradictions inherent in this accommodation.

Conceptually participatory forest management has several important characteristics. It represents a democratic assertion of people's rights; an institutional expression of these rights; and a challenge to the current development paradigm where a demand for rights is not an exclusive pursuit of power but is linked to responsible resource use. The evidence from India and Nepal clearly shows the close relationship between the emergence of democracy as a form of governance and the growing demand for its application to the management of forest resources.

Democratic assertion of rights by forest communities demands decentralised systems. Indeed, given the diversity of cultures, traditions and resource relationships there can be no other effective basis for these systems. In India and Nepal the emergence of local institutions, variously called forest protection committees, village forest committees or forest user groups, is an indication of a movement towards the democratic decentralisation of resource management. The success of these institutions lies in the development of equitable, participatory and responsible relationships.

The task is immense and the prognosis not necessarily favourable. While there is a discernible change in forest policy, there is little evidence to suggest that there is commitment across all interest groups to such a change. The increasing pressure to commercialise forest products may ensure that the relative strength of more powerful interest groups will continue to prevail over decision-making on resources.

The second factor militating against the success of participatory approaches lies in the ownership of these lands by the state – an ownership that is operationalised through the forest departments. These departments have a history, culture and value system based on top-down, centralised processes. The culture of command and values derived from civil and legal procedures that assert the power of the civil servant over that of the forest user contribute to the difficulty of the transition to joint management of resources.

The third factor lies in the complex and heterogeneous structure of villages. Inequities within communities are reflected in highly differentiated resource endowments and power structures. In addition, there are several other social, economic and cultural factors that contribute to the imbalance in power. Attempts to effect structural change through only one sector and resource – forestry and forests – can constitute only one element of a series of changes that need to occur.

Despite these factors, changes are occurring supported by a variety of organisations and individuals, including many innovative forest officers and NGO staff, academics and grassroots organisations. Each contributes to the process of change in a different but fundamentally important manner. Many foresters, for example, are catalysing change from within; on the other hand, there are many people who challenge the existing system from without. Each process is important and they must continue in tandem. There are those who are analysing the details of practice and developing better approaches. Some of these key issues, such as new silvicultural practices, organisational change, equity within and between village institutions, holistic approaches to land-use management, form a major focus for this Study Guide.

The Guide is one of the most comprehensive compilations of experiences gained in participatory forest management in India and Nepal. As such, it should be of use to those who are as yet uninitiated in the process but also to those researchers and practitioners who have been concerned with these issues for sometime. However, the main importance of the Study Guide lies in the fact that it is not only a synthesis of the large and rapidly growing literature on participatory forestry but also a valuable companion to practitioners. It helps both foresters and NGOs to acquire valuable understanding of the historical and current context in which they are operating, and it also provides a guide for future courses of action.

In the historical context of the struggle of forest communities to assert control over their livelihood resources, the Study Guide is an important contribution, which should prove of value to a wide range of actors, be they foresters, NGOs, communities or grassroot activists.

Arvind Khare
World Wide Fund for Nature
New Delhi
India

Mary Hobley's first degree in Forestry was from the University of Wales, Bangor. Her doctorate, based on fieldwork in Nepal investigating the participation of local people in forest management, was awarded by the Australian National University, Canberra. During this time she carried out her research in the Nepal-Australia Forestry Project area.

Whilst completing her doctorate Mary Hobley joined the Overseas Development Institute's Rural Development Forestry Programme (formerly the Social Forestry Programme) in 1987 and worked with Gill Shepherd in the development of the Social Forestry Network. She has conducted research and advisory work mainly in India and Nepal, designing and evaluating new approaches to local forest management. Recent work has included action research into the institutional interface between government and local organisations for forest management, and the development of more effective partnerships through joint forest management projects in India. Since 1994 she has been working with the Karnataka Forest Department in the Western Ghats Forestry Project, funded by the UK Overseas Development Administration, to help in the development of joint forest planning and management techniques.

Acknowledgements

This book has been written for all those concerned with issues of participation in the forest sector: policy-makers, headquarters and field staff of government departments and NGOs; academics; and aid agency staff. The material has been presented to permit its use at undergraduate and postgraduate level as well as in short courses and at workshops.

Responsibility for the views expressed in the text rests with myself and the other contributors, and not with any organisation. I have tried to ensure that interpretation and use of evidence have been careful. However, there are bound to be errors and omissions. I hope that the reader will accept these and that they do not detract from the overall content of the book.

Many have contributed to this book both directly and indirectly. My thanks go to Gill Shepherd for the decade of questioning and challenging we have shared. Some of the experiences discussed in the Guide are drawn from the work of volunteer foresters working for Voluntary Service Overseas. Particular mention should be made of Myriam Dumortier, Ian Whitehead and Stephen Keeling, who have contributed much both to the material used in this book and by their comments drawn from their VSO experience. My very great thanks go to all the rural people who have given me their time and insights over the years. Although some are named in this book, many others have contributed and are nameless. Headquarters and field staff in forest departments and NGOs in India and Nepal have also shared their ideas, concerns and understanding. My special thanks go to Mike Arnold, Jeff Campbell, Barbara Livingstone and Yam Malla for their careful and substantive comments on drafts. In addition, comments and ideas have been contributed by Penny Amerena, Peter Branney, Eric Coull, John Hudson, Stephen Keeling, Arvind Khare, Janet Seeley, Kishore Shah, Dermot Shields and Kaji Shrestha.

My thanks must also go to Mark Agnew for bearing with

patience, kindness and good humour the painful experience of living through the book's gestation period.

My investigations and writing have been made possible through financial and administrative support provided by several organisations. Institutional support was provided by the Overseas Development Institute. My particular thanks go to Peter Gee for undertaking the burden of co-ordinating the production of the book; to Margaret Cornell for her careful and comprehensive editing; and to Ingrid Norton for her support and encouragement. The European Commission provided financial support for research and writing. Publication costs have been provided from the Forestry Research Programme (FRP) component of the UK Overseas Development Administration's Renewable Natural Resources Strategy. I wish to express my gratitude for these grants. Particular thanks go to Howard Wright and Anne Bradley from FRP and to Penny Amerena, Pippa MacBain and Silke Bernau of VSO, for their support, patience and advice in bringing the research and publication to fruition.

In the course of my reading, fieldwork and discussions, I have developed many of my ideas. I recognise that as I write much is changing in India and Nepal and that new events and policies are emerging which will generate new insights. Thus there can be nothing final and there is no last word on any of the issues raised in this book. It is a bringing together of experiences which are pertinent at this moment; perhaps it will stimulate others to pursue new avenues of investigation. However, what is important now is to continue to implement new approaches, to learn and to change in the light of experience.

Mary Hobley

Contributors

Peter Branney: is an independent consultant formerly working with LTS International Ltd based in Edinburgh, specialising in rural development forestry. He has worked in community forestry in Nepal for over three years since 1983, initially as a VSO forester in a remote part of the Mid-Western Region, and more recently as Forest Management Adviser to the Koshi Hills Community Forestry Project based in Dhankuta. Since 1993 he has provided consultancy support to the NUKCFP on participatory forest management. Recently he has started work with the Karnataka Forest Department in the Western Ghats Forestry Project to help in the development of site-specific forest management techniques.

Jeff Campbell: has worked as a programme officer for community forestry with the Ford Foundation for nearly five years. Prior to this he worked as an extension forester, forestry consultant (USAID and FAO), naturalist, wildlife tourism operator, photo journalist, teacher and musician. He has spent over 33 years in India and Nepal.

Yam Malla: is currently co-ordinator and lecturer for the forestry extension courses at the Agricultural Extension and Rural Development Department at the University of Reading. Prior to this he worked with the NACFP as Rural Development Adviser. He has also served as an extension specialist in USAID's Resource Conservation and Utilisation Project, and as Head of the Forestry and Pasture Development Division at the Pakhribas Agriculture Centre in Nepal. For the last two years he has been associated with the Regional Community Forestry Training Centre, Kasetsart University in Bangkok, in the teaching of its community forestry extension courses.

B.M.S. Rathore: is Deputy Director of the Wildlife Institute of India where he is developing approaches to the integration of conservation and development objectives through training and action research in eco-development. He has worked for nearly 10 years as a Divisional Forest Officer in Madhya Pradesh, four years of which (1990–94) were spent in Harda Forest Division working on innovative approaches to participatory resource management.

Verity Smith: is a tutor at the Agricultural Extension and Rural Development Department at the University of Reading, where she is joint coordinator of courses in forestry extension. She lectures on forestry extension at the Regional Community Forestry Training Centre, Kasetsart University in Bangkok.

Eva Wollenberg: is a social scientist at the Centre for International Forestry Research in Bogor, Indonesia, where she is responsible for research on local people's livelihoods, community forests and devolution. Before joining CIFOR in 1994, she was a programme officer with the Ford Foundation's Asia Programme and coordinated a review of the Foundation's support for community forestry work in six countries in Asia.

1 Why Participatory Forestry?

1.1 Introduction

This Study Guide has been written in response to the enormous interest world-wide in the experiences gained in Nepal and India in implementing participatory forms of forest management. It provides a critical analysis of the strengths and weaknesses of these new approaches and considers whether participatory forestry provides a new paradigm for forest management or whether it is another fashionable, but soon to be marginalised, development trend.

The material presented in the Guide draws on the large and growing empirical literature emerging from India and Nepal. Consequently, as well as providing an introduction to the issues, it is also a source book of the information available. Since the literature is so large a bibliography has been included which contains as much of the relevant available literature as it has been possible to collect. Appendices are also included that describe video material and other information sources (Appendices G and H).

Where possible, exercises for use on training courses have also been appended (Appendices A–E). These are ones used both by the author and also by other trainers that have been considerably refined and developed over the course of several years. Discussion points are provided in the margins of each chapter. In some cases these are points on which to reflect; in others they could form the nucleus for a group exercise, or essay questions. A series of additional questions which might be used for group discussions are given in Appendix F. The Guide is not a recipe book of 'how to do' participatory forestry; rather it is a critical analysis of what has happened and what is happening now.

Community forestry in Nepal

1.2 What Are the Origins of Participatory Forestry?

Trends in the forest sector

Throughout much of the last 20 years, international attention has focused on the plight of tropical forests, issues of resource degradation, declining biodiversity and the impact of decreasing forest resources on global climate. As Table 1.1 indicates, the forest sector has adjusted national policies and practices in response to a number of internal and external factors. These factors are discussed in more detail in Chapters 2 and 3 for India and Nepal.

At the international level, proportionately less attention has been focused on local issues of decreasing access to forest resources, and the implications for local people dependent on forests for securing their livelihoods.[1] In recognition of this, local forestry programmes have sought to improve the well-being of forest-dependent villagers. Since government Forest Departments have jurisdiction over public forest lands, an important component of the support of most major international funding agencies has been to promote institutional change within forestry bureaucracies in order to encourage them to be more responsive to the needs of local people. This has included decentralisation of some management control to the local level through a variety of new institutional arrangements, and has also included changes in the policy framework as well as the bureaucratic structure. The following sections examine the emergence of participatory forestry.

Table 1.1 The typology of change

Decade	Event	Response
1970s	• Oil crisis = other energy crisis – firewood • Sahelian drought = deforestation • Bangladesh floods = deforestation	**Forestry for Local Community Development**
1980s	• Eco-disaster = Forestry renaissance	Creation of new forest resources = Woodlots **Social Forestry**
Late 1980s	• Changing development practice = from top-down to bottom-up planning	Local control & management of resources = Participation, acknowledgement of value of indigenous technical knowledge, enhanced role of NGOs
1990s	• New forest sector policies • Rio and Agenda 21 • Decentralisation • Public sector reform	Participatory management = institutional and policy reform, new partnerships **Collaborative, joint, participatory, community forestry**
2000 + Forestry for Multiple Objectives, Multiple Clients, Multiple Partnerships		

The eco-crisis and the basic needs debate

The post-war period from the mid-1940s to the late 1960s was a period of increasing prosperity, rapid industrialisation and full employment within the core countries of the Western world. The economic climate was strongly reflected in modernisation theories which held that poor countries could follow the 'stages of growth' experienced by developed countries if industrialisation and modernisation were stimulated by capital investment (Eisenstadt, 1966; Rostow, 1971). The central concerns of modernisation theory were the dichotomy between 'tradition and modernity' and the assumption that the advance from tradition to modernity is a 'simple unilinear progression' (Higgott, 1978). Aid to the so-called 'Third World' was supplied in the form of large infrastructural packages to develop an economic base from which to promote industrialisation and thus economic development in the expectation of diffusion or 'trickle-down' of benefits to the urban and rural poor (Lerner, 1965).

Modernisation theories permeated all sectors, including forestry. Westoby in a seminal paper of 1962 advanced the argument that industrial forestry would stimulate development in underdeveloped countries (Westoby, 1962). He held that forest-based industries had strong forward and backward linkages with the rest of the economy because they furnished a wide range of goods and services and used mainly local inputs. The demand for forest products was forecast to rise rapidly following the rapid industrialisation of all economies. Douglas (1983) provides a useful critique of the Westoby analysis and contends that the drive to an effective economy can only be achieved through the sound development of a productive rural economy rather than by imposition of a modern industrial framework.

These arguments provided the basis for forest policy development in both developed and less developed countries. They strongly influenced the form of forestry development promoted by the new international aid agencies such as the World Bank and the Food and Agriculture Organisation, among many others (for further references to this era see: Gregory, 1965; Sartorius and Henle, 1968; Keay, 1971; McGregor, 1976; Von Maydell, 1977). At this time in Nepal, working plans were being drawn up for the exploitation of the extensive Tarai sal (*Shorea robusta*) forests. In India too, the increased demand for forest products was met through heavy investment in plantations for the production of industrial wood-based products. Capital was invested in large forest industries supported by the raw material from plantations and intensively managed natural forests (Gadgil *et al.*, 1983).

The boom in Western economies ended abruptly with the economic crises of the early 1970s. Inflation, fuelled by the United States' spending on the Vietnam war, soared further when the OPEC cartel of oil-exporting nations secured a four-fold increase in the price of oil. The economic crises led to a realisation that industrialisation did not necessarily lead to the economic or social development of underdeveloped countries (Griffin and Khan, 1978). Rural and urban poverty became the focus of development theory, with sustenance of 'basic needs' forming the objective of development policy (Streeten and Bucki, 1978; Ghai *et al.*, 1979).

The focus on energy forced attention on the rest of the world where most people are dependent on wood as their main fuel for cooking and heating (Arnold, 1989). A series of reports highlighted the linkages between the millions of people dependent on a rapidly disappearing forest resource and a projected disaster of enormous dimensions (Openshaw, 1974; Earl, 1975; Eckholm, 1975 and 1976; the World Bank, 1978; see also Dewees, 1995 for a critical discussion of the impacts of the 'woodfuel crisis'). At the same time as these concerns were emerging, Frank (1969) was influential in revealing the growing gaps between rich and poor. He showed how the inadequacy of modernisation theories and the policies thus derived from theory had contributed to the increasing poverty of many countries. The debates within development theory pursued the path of fulfilling the basic needs of the poorest and focused on securing the economic advancement of rural populations.[2] This scenario of eco-crisis and livelihood degradation was well developed and has been formative in the construction of forest policy and practice in both India and Nepal.

Forestry, as a follower of development strategies evolved in wider fields, straggled behind the changing moods of development policy. The shift away from industrialisation as the vehicle for development slowly percolated through the forestry sectors of aid agencies. The late 1970s saw a spate of conferences and policy statements. These included Westoby's major rescindment of his

Production and use of tree products at the village level are in practice often embedded in complex resource and social systems, within which most of the factors that affect our ability to intervene with forestry solutions are of a non-forestry nature.
Source: Arnold, 1995

Woman carrying fuelwood, Nepal

1962 paper on the merits of forest industrialisation. He looked back in 1978 at the policies of industrialisation and modernisation that he had so ardently advocated in 1960s and found that '...very, very few of the forest industries that have been established in underdeveloped countries ... have in any way promoted socio-economic development'. At the 1978 Eighth World Forestry Congress ('Forests for People'), where he admitted his disappointment, he elucidated a new social role for forestry, a form of forestry which became known as 'social forestry' and embraced notions of communal action by rural people (Westoby, 1978). The new model to be promoted and followed internationally was stated by FAO to be (note the male orientation):

> Forestry for Local Community Development is a new people-oriented policy . . . the objective of which is to raise the standard of living of the rural dweller, to involve him [sic] in the decision making processes which affect his [sic] very existence and to transform him [sic] into a dynamic citizen capable of contributing to a larger range of activities than he [sic] was used to and of which he [sic] will be the direct beneficiary. Forestry for Local Community Development is therefore about the rural people and for the rural people. (FAO, 1978)

This statement heralded the beginning of a major programme launched by FAO and the Swedish International Development Administration to help the development of community forestry programmes around the world. In the same year, the World Bank issued a *Forestry Sector Policy Paper* which also indicated a major change in direction away from support mainly for industrial forestry to forestry to meet local needs:

> During the last two years, there has been a significant change in . . . Bank activity in preparing forestry projects. Whereas only four of the 17 projects financed between 1953 and 1976 were specifically intended to benefit rural people, over half of the 40 projects in the Bank's forward lending program are people-oriented as opposed to industry-oriented. (World Bank, 1978)

Participatory forestry emerged as a new world-wide practice for forestry development, and was promoted by international organisations and sold in programme and project packages. Although the types of intervention diversified, the profession continued to embrace those traditional practices of forestry which were dominated by the twin dogmas of 'timber primacy and sustained yield' (Glueck, 1986). Forestry was claimed to be the 'unique vehicle' by which the needs of local people could be met and the quality of rural lives enhanced. It was seen as the means by which social change could be affected (Richardson, 1978; Shah, 1975).

Forestry was claimed to be the 'unique vehicle' by which the needs of local people could be met and the quality of rural lives enhanced. It was seen as the means by which social change could be effected.

What evidence is there to indicate that such an outcome is achievable through forestry?

Although much attention was focused on the drudgery and increasing difficulties of fuelwood collection, the social and political problems relating to resource access and property rights were largely ignored. It was naively assumed that increasing physical supplies would provide widely distributed benefits. As Dewees (1995) shows, the size of the presumed fuelwood deficit is not the critical factor; rather it is the quality of the impact on individuals and the amount of labour each household has available to use for fuel collection. As with most issues, the actual response to crisis at the local level is complex and is determined by a number of interlinked issues.

Property rights and participatory forestry

At the heart of participatory forestry lies the battle for ownership of forest lands. Property rights structures have for the last century been skewed in favour of the state, at the expense of local people's needs (see Gordon, 1955). Under recent forestry initiatives, new tenurial arrangements have been introduced. It is not clear, however, that these changes alone have made a sustainable difference in villagers' well-being. In some cases, villagers had *de facto* use rights to forest lands already (and formalisation of these rights has in fact led to a diminution in the benefits available). In other cases, the rights were more short-lived than expected. The history of comminution of rights in Nepal and India is discussed in Chapters 2 and 3, as are the implications for current practice.

Villagers themselves in several countries have raised questions about the security of their claims in the face of political instability and shifting government policies at the national level. Although use rights have been important in increasing villagers' security of access to land, there continues to be debate about whether they should press for full ownership. Advocates of indigenous people's rights feel that these communities should have their original land claims recognised by the state. Such views underpin Principle 22 of the Rio Declaration – a Declaration which guides (or should guide) the approaches of governments to local communities and management of natural resources. The Principle is reproduced here as it describes the new 'philosophy' and provides the ideological backbone for interventions in the forestry sector.

> Indigenous people and their communities, and other local communities, have a vital role in environmental management and development because of their knowledge and traditional practices. States should recognise and duly support their identity, culture and interests and enable their effective participation in the achievement of sustainable development.

In the following sections, the background to the development of participatory forestry approaches in South Asia is considered, including an analysis of the global context in which policies of decentralisation and divestment of public sector authority have become the currency of action. This provides the context in which to consider in detail the major defining features of the policy and implementational differences between India and Nepal.

What is participation?

The panoply of terms spawned by new development interventions requires careful assessment and use, since they have as many meanings as there are users (Picciotto, 1992). This is particularly the case with 'participation' which over the last 10 years has become one of the most widely used words in the development dictionary. However, as Table 1.2 indicates, what is meant by participation is highly context-specific and its effects range from coercion to full local control. The cynical view of participation lies at the coercive end of this continuum, and will be considered further in Chapters 4 and 5.

In each of the case studies presented these aspects of participation will be considered. At root many of the problems currently being experienced in both India and Nepal can be traced back to

Table 1.2 A typology of participation: how people participate in development programmes and projects

Typology	Characteristics of each type
1. Manipulative Participation	Participation is a pretence with people's representatives on official boards but unelected and having no power
2. Passive Participation	People participate by being told what has been decided or has already happened. It involves unilateral announcements by an administration or project management without listening to people's responses. The information being shared belongs only to external professionals.
3. Participation by Consultation	People participate by being consulted or by answering questions. External agents define problems and information-gathering processes, and so control analysis. Such a consultative process does not concede any share in decision-making, and professionals are under no obligation to take on board people's views.
4. Participation for Material Incentives	People participate by contributing resources, for example labour, in return for food, cash or other material incentives. Farmers may provide the fields and labour, but are involved in neither experimentation nor the process of learning. This is commonly called participation, yet people have no stake in prolonging technologies or practices when the incentives end.
5. Functional Participation	Participation seen by external agencies as a means to achieve project goals, especially reduced costs. People may participate by forming groups to meet predetermined objectives related to the project. Such involvement may be interactive and involve shared decision-making, but tends to arise only after major decisions have already been made by external agents. At worst, local people may still only be coopted to serve external goals.
6. Interactive Participation	People participate in joint analysis, development of action plans and formation or strengthening of local institutions. Participation is seen as a right, not just the means to achieve project goals. The process involves interdisciplinary methodologies that seek multiple perspectives and make use of systemic and structured learning processes. As groups take control over local decisions and determine how available resources are used, so they have a stake in maintaining structures or practices.
7. Self-Mobilisation	People participate by taking initiatives independently of external institutions to change systems. They develop contacts with external institutions for resources and technical advice they need, but retain control over how resources are used. Self-mobilisation can spread if governments and NGOs provide an enabling framework of support. Such self-initiated mobilisation may or may not challenge existing distributions of wealth and power.

Source: Bass *et al.*, 1995

the form of participatory practice developed by the project or programme. Criteria developed for assessing the effectiveness of participatory processes and of the local organisation formed as a result are described in Chapter 4, accompanied by case-study material to illustrate some of the major points.

1.3 The Decentralisation Debate in the Forestry Sector

Why has participatory forestry become such an important initiative within the forest sector? One of the major reasons results from the desire of the international community to achieve sustainability and efficiency through decentralisation and public sector reform. Participatory forestry represents the major attempt to achieve this aim. The new management ethos talks about clients, stakeholders and interest groups, and asks the private and public sector to identify their client groups and their needs, and to respond with services that will support these groups. This new managerialism is mirrored by political theory, where decentralisation also requires clear identification of stakeholders, placing control and authority with these groups, with government bureaucracies restructuring to support their clients. The institutional change implied by these approaches is far-reaching.

Elements of these changes are still unexplored within the forest sector, although, as is discussed in Chapter 7, forestry projects charged with facilitating institutional change are now beginning to address these issues.

Forest Departments, in common with other government agencies across the world, are facing hard questioning concerning their future role in the sector. In New Zealand, for example, government took the radical step of privatising the Forestry Commission. In the UK the form of forest sector management is still to be decided, but undoubtedly there will be some change, as indicated by the split between a forest authority and a forest enterprise. In the USA, the Gore Report (1993) has had equally far-reaching impacts on the domestic forest service and also on the agency charged with overseas development. In India, public sector reform is emerging onto the public arena, prompted in part by the actions of the World Bank.

This Study Guide considers the following questions surrounding the impact of decentralisation as it is manifested through participatory management practices within the forestry sector:

- What are the impacts of this process on the formal and non-formal forestry institutions?

- Under what new institutional arrangements should forests be managed?

- How central is a restructuring of the property rights framework to enable effective decentralisation?

- Who are the winners and losers?

Decentralisation versus devolution

There are many questions still to be addressed about the effectiveness of decentralisation as a political tool to ensure devolution of power, as Webster (1990) indicates:

> Decentralisation has been seen as a means by which the state can be made more responsive, more adaptable, to regional and local needs than is the case with a concentration of administrative power and responsibility in the central state ... But decentralisation of government in itself does not necessarily involve a devolution of power. The extension of the state outwards and downwards can equally serve the objective of consolidating the power of a state at the centre as well as that of devolving power away from the central state; it can both extend the state's control over people as well as the people's control over the state and its activities. Decentralisation is a two-edged sword.

The penetration of the state and centralisation of control are discussed in detail, in Chapter 4, with respect to the development of local-level organisations.

Although the calls for devolution of power to the local level are pervasive across the international community, and all recognise the central role of local users of resources in management, how effective has this devolution been? As Webster (1990) indicates, is it necessarily such a 'good thing'? Since much of the experience gained with the implementation of new forms of forestry is relatively recent, it is perhaps too soon to be able to pronounce definitively on success or otherwise. However, early indications, as discussed in this book, do indicate that rhetoric and reality remain far apart. Thus although major donor organisations and international agreements may all subscribe to the following view, the reality of such a goal is still distant:

> The pursuit of sustainable development requires a political system that secures effective participation in decision-making ... This is best secured by decentralising the management of resources upon which local communities depend, and giving these communities an effective say over the use of these resources. It will also require promoting citizens' initiatives, empowering people's organisations, and strengthening local democracy. (WCED, 1987 cited in Colchester, 1994)

The World Bank, FAO through its Tropical Forest Action Programme, ITTO and IUCN all share this central tenet of local participation in the management of resources (see recent policy documents such as the World Bank, 1991a and 1994). The extent to which such principles can and should direct development policy in the forestry sector is still to be questioned. At the root of this rhetoric is there a real quest for a new world order where actions are assessed in the light of their impact on individuals, and where governments and their agents are held accountable at the most local level? Some would contend that this should be the underlying thrust of the approach (Ghai, 1994; Colchester, 1994); others see it as a means through which to decrease the costs of government, and enhance the participation of the private and other sectors (World Bank, 1992 quoted in Roychowdhury, 1994). Is it a call for a new democratic structure that allows those at the local level control over their destinies (see also Colchester, 1994)? Furthermore, is forestry an appropriate vehicle through which to challenge the existing form of governance?

'The predominant (development) approach pursued in developing countries has been characterised by excessive centralisation, large-scale investment and modern technology, and has often resulted in sharp inequalities and widespread impoverishment. It has frequently been environmentally destructive and socially disruptive, with unregulated industry and concessions to capitalist interest contributing both to environmental degradation and the dispossession and impoverishment of indigenous people. The alternative approach to development, which is exemplified by the grassroots environmental movements . . . is characterised by small-scale activities, improved technology, local control of resources, widespread economic and social participation and environmental conservation'
(Ghai and Vivian, 1992).

Do decentralisation and devolution lead to greater equity? Is this an obtainable goal? Is the obverse of this centralisation and inequity? Is the quest, spearheaded by Western-based doctrines, for efficiency, accountability of public organisations, divestment, privatisation, an appropriate response to the needs of villagers wanting to gain greater control over the use of and access to natural resources? Some influential commentators on the political economy of countries such as India, question the validity of a direct transfer of Western ideology (Ghosh, 1994). This chapter, and indeed the Study Guide as a whole, does not attempt to answer these questions but tries to assemble some evidence to indicate the complex nature of the impacts of decentralisation (whether partial or total), and thus the difficulties and successes of developing participatory forestry approaches that can match the complexity of environments in which it is being developed.

What type of partnership?

The arguments surrounding the decentralisation debate involve discussion of what is an appropriate institutional form to manage forest resources. As the following sections indicate, there is no one solution to this question, but rather an array of arrangements according to the particular requirements of the forest users. How far the forest bureaucracy can or will divest itself of some of its authority remains to be seen. However, in an atmosphere of increasing intolerance of bureaucratic ineptitude, there seems little doubt that forest services will be forced to divest some of their authority, at least at the margins of their power base, with the release of some degraded lands to joint management schemes with local people.

Just as questions are being asked about the role of the state in regulation and management of natural resources, so too are questions being asked about the nature of local organisations being developed by governments and the interests of those they represent (Colchester, 1994; Hirsch, 1993). Policies in Thailand that have encouraged the penetration of the state into regions previously managed by indigenous institutions have produced questionable benefits for the majority of local people: 'These "participatory institutions" which purportedly give the village a role in making rural development decisions are the facilitators of a paralysing bureaucratisation of village procedure which has replaced the older more informal institutions' (Hirsch, 1993). Westoby (1987), reflecting on community development practices of the 1960s and 70s, echoes these sentiments:

> Only very much later did it dawn on the development establishment that the very act of establishing new institutions often meant the weakening, even the destruction, of existing indigenous institutions which ought to have served as the basis for sane and durable development: the family, the clan, the tribe, the village, sundry mutual aid organisations, peasant associations, rural trade unions, marketing and distribution systems and so on.

It is disingenuous to characterise development as the two simple alternatives – decentralisation or centralisation, local people versus government. Together with the contention that grassroots environmental movements are necessarily going to lead to more widespread benefits, this has to be carefully evaluated.

The call for grassroots development brings into question the conditions under which it is appropriate. As the vast literature on collective action shows (Wade, 1988; Ostrom, 1990; Bromley, 1992; Ostrom, 1994), there are many conditions under which collective action has broken down and resources have degraded. The defining features under which such action is appropriate remain elusive in the forest sector, although certain patterns are emerging – most particularly those seen in resource-scarce situations, well illustrated in the Middle Hills of Nepal (see also examples in the African rangelands (Runge, 1986; Shepherd, 1992). These conditions will be considered in greater detail in Chapter 4.

What is forestry?

Although this may seem to be a trite question its answer provides many of the reasons why decentralisation and the role of participatory forestry have become such important and all-pervasive questions in the forestry sector.

Forestry encompasses many objectives: commercial, rural development (poverty alleviation, employment creation, empowerment of marginalised groups – in particular, women), tourism and amenity, and conservation. Conflicts often arise between these objectives and the priority assigned to each in a given area. Research disciplines required for the support of forestry include: economics, micro-biology, history, increasingly political science, anthropology, sociology, law, ecology, chemistry (soil science), zoology, botany, among many others. Forestry, alone among the professional disciplines, derives its power base from ownership of large areas of land. It is highly centralised with a diversity of roles and products, where internal conflicts and contradictions often dominate. Its practice has required the development of multi-disciplinary skills and their accommodation within a framework that allows their full expression. The power base derived from its landholdings has also left it vulnerable to attack by a number of environmental and human rights groups who contend that this power has been wrongfully wrested from those local groups whose livelihoods are deeply associated with the forests.

Timber, logging concessions, government officials, local forest users, democratic institutions, corruption – all these words link up in different forms of open and hidden relationships. As early as 1975, Jack Westoby, reflecting on 20 years of development assistance to the forest sector questioned its contribution to the economic and social life of underdeveloped nations. Still, in many countries of South-East Asia, the nexus between timber, the state, and the trade is seriously undermining the development of any form of local democratic institution for the management of forest resources: 'The practice of dealing out logging licences to members of the state legislature to secure their allegiance is so commonplace in Sarawak that it has created a whole class of instant millionaires' (Colchester, 1989). The conflicts between macro-political need, donor imperatives, and local needs are becoming increasingly more clearly articulated as participatory forestry and decentralisation become common currency. Thus the potential impact of decentralisation on the formal and informal institutions is dramatic. As Forest Departments have been forced through economic and political expediency to adjust their structures certain features of these institutions have become more apparent. These features are considered in more detail in Chapter 7.

From the catapulting of forestry on to the international stage, in the early 1980s, to the grassroots questioning of the role of the profession, the response has been a defensive one of seeking new forms of partnership that will help to deflect some of this criticism (see Chapter 4). Together with the global climate of decentralisation and bureaucratic divestment, this has led to the current situation where forestry (so long impervious to the decrees of the outside world) has been forced to respond to these changes and examine its own institutional framework. This framework now contains responsibility for a wide range of often conflicting land management objectives, as indicated above. Structures that were established to fulfil the primary objective of revenue maximisation are now redundant in a world that insists that forest lands be managed for a multiplicity of benefits. The emergence of a new silviculture in response to multiple objective management is discussed in Chapter 6.

The change from a primary objective of revenue maximisation to multiple objectives ranging from conservation management to development of local organisations for forest management has had profound consequences across the forestry sector. The debate about decentralisation is by no means confined to the developing world but is live in every country.[3]

The implementation of decentralisation processes has brought issues of ownership and control to the forefront of debate. In forestry, the historical development of state control over forest lands has meant that the land base held in trust by the institution for the public good is enormous. The following statistics provide an indication of the extent of forestry estates in Asia. In India, Forest Departments control 22% of the national territory (Agarwal and Narain, 1989); in Nepal forests and shrublands comprise some 43% of the total land area (Nield, 1985); in Indonesia, 74% of the territory is controlled by the Forest Department; and in Thailand, the Royal Forest Department administers some 40% of the nation's land (Colchester, 1994). These extraordinary figures underline the fundamental challenge posed to these departments by the call for devolution of some of this control to the millions of people living in forest areas. The means by which this is being done needs considerably more analysis and the form of the linkages between state and people needs to be critically assessed (see Chapters 4 and 5 for further discussion).

Decentralisation in action

At one extreme of the public to private sector continuum lies the New Zealand Forest Department, where probably one of the most far-reaching restructurings of the sector has occurred. Here, the forest service was abolished and separate organisational structures were established. This deconstruction of a monolithic organisation in favour of several discretely functioning units has been one mechanism to cope with the conflicts of multiple objective management engendered within one organisation. This is described in a statement in the 1987 Report of the Director-General of Forests:

> The major reasons which led to the restructuring of the New Zealand Forest Service were an inability to provide the transparent accountability for the mix of functions performed by the department and perceived conflicts of interest between those functions. (cited in Brown and Valentine, 1994)

By identifying and separating out these objectives and forming distinct organisations each with primary responsibility for a major objective, conflicts become public (i.e. inter-departmental wrangling is more visible than intra-departmental disputes). Such an approach is also recommended for India by Gadgil and Guha (1992b). Demarcation of territorial responsibility and therefore also accountability is easier to attribute. As such, the advisory and regulatory functions are the responsibility of a Ministry of Forestry. Conservation, a subject which has frequently brought forestry professionals into conflict with environmentalists, and which is considered by many to be irreconcilable with the practice of commercial forestry, has been assigned to a Department of Conservation (primarily responsible for natural forest conservation). The state-owned Forestry Corporation was made responsible for commercial, plantation resource-based forestry activities. In addition, the great power base of a forest service – its land – has also been largely privatised.

The strong message that emerges from the New Zealand experience is that there is no blue-print for institutional change: the structure of organisations necessary to meet international, national and local imperatives must emerge from the particular circumstances of each nation. The principle of decentralisation, although global, does not necessarily lead to a globally uniform response. These responses are discussed in detail in Chapter 7, where the implications of the transition from public to private sector operation and the degree to which divestment can and should occur are assessed for Nepal and India.

1.4 The Development Forestry Context: South Asia

South Asia has been witness to a series of dramatic experiments in the participatory management of forest resources. Since the 1970s social and community forestry programmes in both India and Nepal have attempted to transform the relationship between a powerful state bureaucracy and local people directly dependent on forest resources. These programmes represent the realisation that a large proportion of the population depends heavily on forest

A social forestry plantation, Bihar, India

Box 1.1 Terminology for Participatory Forestry

Social Forestry

In the Indian context, social forestry refers to Forest Department-sponsored plantations on a variety of 'wastelands', such as village grazing commons, government-owned revenue lands, roadsides and canal and tank banks undertaken with varying degrees of local participation. The term was first used by the Indian Government during the 1970s, as a land-tenure term for forestry on village, and not forest reserve, land. In 1978, the term was used by Westoby at the World Forestry Congress to mean 'forestry for local needs'. In the 1980s it became an umbrella term for individual farm forestry, for communal village planting and in some places for forest management by villagers.

Farm Forestry

Farm forestry involves the promotion of tree planting by farmers on private lands through free and low-priced seedlings and decentralised nurseries.

Community Forestry

Community forestry is a broad term which includes indigenous forest management systems and government-initiated programmes like user group forestry in Nepal and joint forest management in India, in which specific community forest users protect and manage state forests in some form of partnership with the government.

Although community forestry is most closely associated with Nepal, as indicated above it is widely used by international agencies to describe people-based forms of forest management. The significant difference between India and Nepal lies in the nature of the benefit-sharing arrangement. Local people in Nepal are allowed 100% of the benefits flowing from a forest under their management, with the provision that a percentage of the benefits are directed back into enhancement of the forest resource.

Joint Forest Management (JFM)

JFM of forest lands is the sharing of products, responsibilities, control, and decision-making authority over forest lands between Forest Departments and local user groups. It involves a contract specifying the distribution of authority, responsibility, and benefits between villages and State Forest Departments with respect to lands allocated to JFM. The primary purpose of JFM is to create conditions at the local level that enable improvements in forest conditions and productivity. A second goal is to support a more equitable distribution of forest products than is currently the case in most areas (Moench, 1990).

Rural Development Forestry (RDF)

RDF is the growth and management of trees where primary management decisions are made by users of the trees, either as individuals or groups, and where the primary benefits of trees remain within the household or community (Warren, 1992).

resources for subsistence, energy, nutrition, income and the maintenance of farming systems. They acknowledge the failure of traditional custodial management of forests by government to halt the loss and degradation of the sub-continent's forests, without the active participation of local communities.

The inadequacy of government-based approaches to forest protection and management led to the search for alternatives, and experimentation with a number of approaches. These can generally be classified into social forestry, farm forestry, community forestry, joint forest management and rural development forestry (see Box 1.1). In this Guide, the umbrella term used to refer to all these approaches is participatory forestry, accepting the diversity of interpretations of participatory. Although, as some have contended (Fisher, 1995; Johnson, 1995), the use of the word 'participatory' is probably more problematic than some of the more clearly focused terms such as 'collaborative' or, as Johnson suggests, 'good forest management', it is used here, however, because the breadth of interpretation associated with it is one of the main characteristics that this Study Guide explores.

The earliest mention of 'social' in forestry was in India where several States pioneered tree-growing programmes outside the traditional forest boundaries (Gadgil *et al.*, 1983; Wiersum, 1986; Arnold, 1989; see also Box 1.2). For example, the State of Gujarat in 1970 set up a Community Forestry Wing in the Forest Department, and Tamil Nadu started a tree-planting programme for local employment generation on tank foreshores and village wastelands as early as 1956. After 1973 half of the proceeds from these plantations were given to local panchayats (the lowest unit of local government administration) and local people were allowed to collect fodder from the plantation areas (Wiersum, 1986). Under some interpretations of social forestry it could be considered that its formal origins lie in government programmes of the late nineteenth century where 'village forests' were demarcated (Pardo, 1985). However, under other interpretations this would be considered to have been a

Box 1.2 Why Is 'Social' Important in Forestry?

- The adjective 'social' is used in a descriptive way; it indicates public involvement in forest management. Such participation is mainly seen as a means to achieve the objective of effective forest protection.
- The adjective 'social' is used in a normative way. It indicates a social development norm: the objective of the forestry scheme is directed at fulfilling human needs. Public participation is the main objective of many such schemes, and forest (or tree) management is the vehicle by which to achieve this objective.
- The emergence of 'social forestry' has tended to reinforce the tendency to treat it as a separate programme area definably different and separate from existing programmes, such as forestry, obscuring the need to revise 'forestry' to incorporate the additional dimension of meeting local as well as national and industrial needs.

Source: Wiersum, 1986; Arnold, 1989

programme of removal of local people's rights to manage forests (Guha, 1983). Indeed, many commentators in both India and Nepal would assert that participatory forestry has been implemented, informally and unrecognised, by local people over many decades and generations, and that the so-called 'new' approaches are merely reproducing (often badly) indigenously derived systems of forest management (Fisher, 1991; Bartlett and Malla, 1992; Hobley *et al.*, 1994).

Thus, by the early to mid-1980s it was possible to make some assessments of these social and community forestry programmes which had been running for over a decade (Arnold, 1989). The dichotomy of understanding the meaning of 'social' in social forestry has interesting and long-running consequences for participatory forestry. In the early years of external funding of forestry projects, the justification for the funding was given on the basis of poverty alleviation, where forestry was seen to be the appropriate entry point to reach the more marginalised groups in society. However, as evidence from India indicates, this ideal was far from realised through the social forestry programmes, and in many instances poorer groups were dispossessed from the land they had been using, particularly those groups whose livelihoods were dependent on access to grazing lands. The mix of objectives ascribed to social forestry doomed the programme to difficulties from the outset, with a multiplicity of target groups to be reached but only one model – that of woodlots.

Planting trees to meet environmental objectives such as soil protection is unlikely to produce sufficient output of saleable products to be economically attractive to the farmer. Similarly, tree growing designed to generate income is unlikely to benefit those with little or no land. Production to meet both subsistence and market needs is unlikely to be achieved with a single production model. Projects originally designed to meet a production goal are unlikely to be equally successful at achieving a subsequently added social goal, such as favouring the poor.

(Arnold, 1989)

Although there is evidence to indicate that farm forestry in certain parts of India proved to be immensely successful in the initial stages, as demonstrated by the demand for seedlings which far outpaced projects or supply, private tree growing on a large scale was confined to parts of North-western India, Gujarat and Karnataka, resulting in localised over-production of poles and a consequent depression in prices. Perhaps because of falling prices and local surpluses, the initial boom in farm forestry has slowed (Saxena, 1994a).

Reviews of social forestry programmes, which had objectives of developing the common property resource, have been far less positive. One of the common factors identified in their failure was the absence of people's participation in planning and management, which led to poor survival rates and the reluctance of community institutions to take over responsibility for the management of plantations. Furthermore, even though both these programmes shared the common objective of reducing pressure on forest lands through creating alternative sources of fuel, fodder and forest products, degradation still continued. The intense focus of funds and energy on private and common lands in India redirected attention away from investment and management of natural forests (Arnold *et al.*, 1987a & b; Arnold, 1990; Chambers *et al.*, 1989; Chaffey *et al.*, 1992).

It is this background that led to the emergence of a fundamentally new practice – community forestry in Nepal by local people, or joint forest management, as it is known in India, involving local people actively in the protection and management of state forest lands. While community forests are being managed in Nepal, joint forest management arrangements are being explored in India between local people and State Forest Departments. In the process many self-initiated and indigenous forest management systems are being documented and are gaining recognition (Arnold and Campbell, 1986, King *et al.*, 1990, Gilmour, 1990; Gilmour and Fisher, 1991, Campbell and Denholm, 1993, Chhetri and Pandey, 1992, Karki *et al.*, 1994, Kant *et al.*, 1991). Social forestry and farm forestry were the first new practices in recent history to bring foresters out of the forest and into the villages and farms of the people who are the forests' primary users. New community forestry programmes seek to go a step further, recognising the role of these users in the management of natural forests – bringing the people back into the forests.

1.5 Discussion

In a workshop to exchange experience between practitioners of social and community forestry in India and Nepal the outcome suggested that, although there were many similarities in experiences, there were also some major differences (Campbell and Denholm, 1993). In many cases, failures in one country were mirrored at a later date in the other, indicating that although these two nations may have many points of interaction there had been little or no sharing of experiences in the forestry sector. It is estimated that over $2 billion has been invested by donors alone on these programmes over the last 15 years. National and state Forest Departments are now allocating or re-directing substantial funds, often with large donor assistance, for community/joint management. Yet these new forestry experiments are still evolving, and their focus on local institutions and equity make them more process-oriented, and less amenable to rigid target-based development planning. People's participation, reorientation and training of forest staff, building local-level institutions, participatory microplanning, equitable benefit sharing, gender-sensitive programming have all become new development imperatives (Arora, 1994). Community forestry in Nepal and joint forest management in India are beginning to take on these challenges in different ways.

The essence of current changes in forest management in both Nepal and India lies in the attempt to shift control and management of forest land from centralised Forest Departments to decentralised people's organisations. The historical background and legal basis to the two programmes are unique to each country, although they do share certain similarities (particularly in recent years with the new 'hegemony' of aid programmes (see also Leslie, 1985 and 1987)). The types of community institutions, though they are still evolving and share many features, are distinct and differ between countries and within States in India. The nature and extent of the shift of control from State/national to local/community level also differs considerably. It is in the implementation at various levels that a greater degree of overlap exists, although the sequence of planning and ownership of management vary significantly. Ironically, the programmes in both countries have focused more attention on initiating community protection (India) or simple operational plans (Nepal) than on making the more dramatic shift to active co-operative forest management and to addressing the technical, social and economic issues which accompany such a transition. Many of the problems, faced by both countries, are therefore very similar. Chapters 2 and 3 look at the historical factors leading to the emergence of participatory forms of forest management in India and Nepal. A summary table (Table 3.3) in Chapter 3 reviews the similarities and differences between the two countries. Annex 1.1 describes the physical aspects of the resources in India and Nepal.

Annex 1.1

The Resource Base in India and Nepal

India (see Figure 1.1)

The total forest area recorded is about 77 million hectares which constitutes about 23% of the country's total geographical area (329 million ha). The diversity of eco-zones in India is only matched by the diversity of social systems and it is impossible to describe either fully in a book of this nature. Instead, description is limited to a discussion of the major forest types, and where appropriate the particular ecological and social conditions of a case-study area will be provided. Box 1.3 provides a summary description of the major forest types in India.

Figure 1.1
India

Box 1.3 Forests of India

There are four major classifications of forest types following the temperature-based climatic zones: tropical, sub-tropical, temperate and alpine. These have been further sub-divided into 16 sub-groups. Of the, 5 sub-groups are of particular significance due to their extent:

- Tropical moist deciduous forest (37%)
- Tropical dry deciduous forest (29%)
- Tropical wet evergreen forest (8%)
- Tropical semi-evergreen forest (8%)
- Himalayan dry temperate forest (7%)

Other types include:

- Montane wet temperate forest (4%)
- Himalayan temperate forest (0.05%)

Major tree species include:

- Dipterocarps, sal (*Shorea robusta*), teak (*Tectonia grandis*), sissoo (*Dalbergia sissoo*), laurel (*Terminalia* sp), bijasal (*Pterocarpus marsupium*), bamboo (*Dendrocalamus* sp, *Bambusa* sp and several other varieties), *Anogeissus*, *Gmelina*, *Albizzia* sp, kendu (*Diospyros melanoxylon*) and mahua (*Madhuca latifolia*)

Source: Johri and Babu, 1974; Femconsult, 1995a

Nepal

Nepal now has over 5.5 million hectares of natural forests which equates to 37% of its land area. Only 11% of the natural forests are in the Tarai and high Himal zones; the remaining area is evenly distributed across the Middle Hills and the Siwaliks (see Figure 1.2). Of this land area 61% has been identified as potential community forests – forests which could be handed over to local people for management. Much of the forest in the Middle Hills is found in small patches surrounded by cultivation; there are few large tracts of forests amenable to conventional forms of forest management. It is here where there is extreme pressure on forests and where livelihoods are most intimately associated with forests. However, what is interesting is that, although crown densities may be decreasing in some parts, though not in all (Gilmour and Nurse, 1991), the actual area of land under forests does not appear to have changed significantly. Indeed in some areas there is actually an increase in land under trees (Carter and Gilmour, 1989; Carter, 1992).

Box 1.4 Nepal's Forest Types

Natural resource management systems in Nepal reflect both the country's long and varied political history and the prevailing conditions in its five major physiographic regions, each of which occupies a horizontal band that stretches across of the country from east to west. Altitudinal variation is extreme and impacts on ecological conditions and climate.

Nepal's southern lowlands are known as the Tarai and are a subtropical extension of the Gangetic Plain. Physical relief varies less than 1%. This area of extremely fertile land accounts for 60% of the country's total agricultural output. With a relatively well developed infrastructure and easy access to the markets of northern India, the Tarai also serves as the country's industrial centre. Although it accounts for only 14% of the land area, the Tarai is now home to about 45% of the population. This area contains some remnants of the tropical moist deciduous forest, and the much depleted sal forests.

Immediately north of the Tarai are the Churia Hills (Siwaliks) – relatively low, parallel ridges that run the length of the country and enclose several elongated valleys known as 'duns'. Ranging in elevation from 120 metres in the east to nearly 2,000 metres in the far west, the Churia Hills account for 13% of the total land area. With steep slopes and mostly poor, shallow soils, the Churia Hills are not well suited to cultivation, a condition which has 'protected' about 26% of the remaining natural forests in Nepal.

The Middle Hills are characterised by the poor state of the forests. These areas account for a third of the total land area but accommodate nearly half the population. Elevations in the Middle Hills range from 200 metres to over 3,000 metres in the Mahabharat Lekh, the major foothill range of the Himalayas. Despite widespread deforestation, over one-third of Nepal's forests are found in this area. This area has a temperate monsoonal climate supporting mainly rainfed terraced agriculture with some irrigated agriculture in the valley bottoms.

The fourth region of Nepal is the High Mountains. The upper boundary of this area corresponds with the tree line at about 4,200 metres, while the lower boundary varies between 1,000 metres on the valley floors and 3,000 metres on the ridges. The High Mountains contain about 30% of the natural forests on about 20% of the total land area.

The main forest types of Nepal described by Stainton (1972) follow closely Champion's classification of India's forest types (given in Box 1.3 above).

- Tropical and sub-tropical (sal forest – major proportion)
- *Terminalia forest (Dalbergia sissoo-Acacia catechu forest)*
- Sub-tropical deciduous hill forest (*Schima-Castanopsis forest)*
- Sub-tropical and semi-evergreen hill forest (*Pinus roxburghii forest)*
- Temperate and alpine broadleaved forest (*Quercus, Castanopsis, Aesculus, Juglans, Acer forests, etc.)*
- Lower and upper temperate mixed broadleaved forest (*Rhododendron and Betula forests)*
- Temperate and alpine conifer (*Abies, Cedrus, Pinus, Tsuga, Cupressus, Larix forests)*
- Moist alpine scrub
- Dry alpine scrub

Source: Stainton, 1972; Talbott and Khadka, 1994

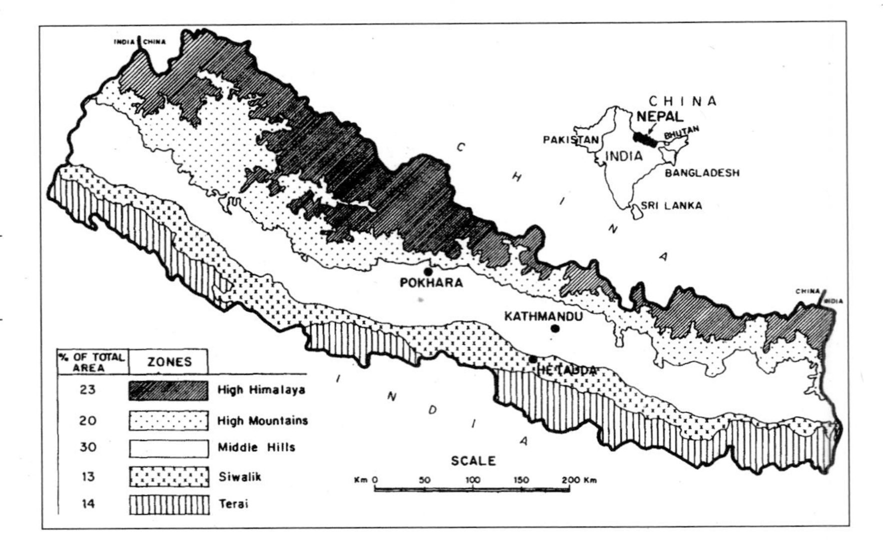

Figure 1.2
Nepal

Notes

1. Livelihoods are described as 'an adequate and secure stock and flow of cash and food for the household and its members throughout the year' (Chambers *et al.*, 1989).
2. Detailed discussion of the history and development of these theories lies outside the scope of this book. Leys (1977) argues that adoption by the World Bank of its 'poverty-oriented aid philosophy' under McNamara's presidency, and the corresponding 'reorientation of bilateral aid doctrines by the USA and other countries' indicate the problems inherent in underdevelopment theories (i.e. they can be coopted). As a result of this development in theory and the change to a 'needs-based' focus, the World Bank and other international agencies restructured their aid activities towards the promotion of rural development, which was defined as a 'strategy designed to improve the economic and social life of a specific group of people – the rural poor' (Harriss, 1982). For a discussion of the relationships between the forest sector and development theory see Douglas, 1983. For a useful critical analysis of the role of foreign aid see Riddell, 1987.
3. See a recent edition of *Unasylva* (vol. 45(178), 1994) devoted to a discussion of the impacts of decentralisation on the forestry sector.

2 The Four Ages of Indian Forestry: Colonialism, Commercialism, Conservation, Collaboration

...the unpopularity of the Forest Department is a widely recognised reality and it is useless to ignore it (Anon, 1885)

2.1 Introduction

In contrast to Nepal where government control of the hill forests was only ever *de facto* and never really *de jure*, 95% of India's forest land is owned and managed by State government Forest Departments (Singh, 1990). India's forest estate has been under extensive management, even in many of the most remote areas, for the last 100 years. Since its inception in 1864 the presence of Indian Forest Service officers and State Forest Department field staff has been continuous in most of the forest area, and even though the major function may have been custodial, the separation of government forest lands from community lands was complete in the minds of both the local communities and the government. The contentious nature of the relationship between the Forest Department and the rest of civil society is considered in the context of its 130-year history.

This chapter considers the historical evolution of forestry practice in India and follows its stages of development through distinct if often overlapping phases. These phases start with the development of forestry practice under the British during the colonial period and move through periods of commercialism, increased awareness of the need for conservation and finally to the present day. This current phase is marked by an awareness of increasing complexity where objectives of conservation, support to local livelihoods and supply of industrial requirements are melded together in collaborative approaches.

2.2 The Age of Colonialism[1]

The situation today is the result of a series of laws and policies evolved over more than a century beginning during the colonial period of British rule, which have nationalised community and private forest lands and gradually eroded the rights and concessions of surrounding forest communities. Thus an understanding of the historical context is essential in order to interpret current reality. Complex layers of rights, concessions, powers and duties underlie Indian forestry law and are discussed in this chapter.

The long history of development of forestry within India has several interesting dimensions which are not only relevant to debate in India today but are also of consequence to many other forest sectors around the world. Much of what constitutes forestry in anglophone areas is based on the experiences in India. For example, India's forest policy was used as the model for other colonial countries and thus many similarities can be seen (Gordon, 1955).

As early as 1815 the Bombay Government established the first 'rules for the management and preservation of forests in the provinces of Malabar and Canara and to define the authority of the Conservator' (Buchy, in press). The objectives of the East India Company were two-fold:

- to preserve the Imperial (British) forests from waste; and
- to uphold the Company's right over the timber resources.

From the beginning of the nineteenth century, British officials had encouraged the conversion of forest lands into agricultural and hence revenue-paying lands. By the middle of the century the economy was being reoriented towards colonial interests with a massive development of infrastructure and increased pressure on decreasing forest resources. The domestic market demanded timber as the urban population grew, particularly in the port cities of Bombay and Calcutta, and the government demanded timber to build military cantonments and public works. There was also a growing market for teak for both export and ship-building. Not only were the extensive sal (*Shorea robusta*) forests in the drier areas cleared, but the valuable forests of deodar (*Cedrus deodara*) in the Himalayas and teak in South India were devastated. In 1830, the India Navy Board counselled reintroduction of a forest conservation policy, following reports that foresaw the disappearance of the forests within a few years.

The colonial response to the forest problem was constructed from three separate perceptions of forest management. The first was the direct response of military engineers bringing the inefficient logging of some Himalayan forests under their own direction in order to secure government supplies (Tucker, 1986a). The second resulted from the attempts of some officials to conserve forests for catchment protection and to protect them from becoming 'wasteland' – a concept widely used in colonial forestry, and discussed below. (This approach is well exemplified by the work of Dr Cleghorn, an assistant military surgeon in Madras, who in 1861 wrote an important book *The Forests and Gardens of South India* and was deeply concerned with the conservation of the environment.) The third perception of appropriate forest management was provided by those senior colonial administrators who had experience of forestry as it related to wildlife management on their estates in Britain. This rather diverse array of conceptions and enthusiasms led to a disparate series of policy debates and to some notable conflicts.

In 1847, the first attempt to regulate the exploitation of forests was begun with the appointment of two officials given authority to conserve forests. In 1851, the British Association for the Advancement of Science met in London to discuss the serious nature of forest devastation in India, and reported to Parliament on this matter. As a result of these deliberations, in 1855 the Viceroy, Lord Dalhousie, declared a forest policy and appointed the first professional forester, Dr Dietrich Brandis, to superintend the exploitation of the Burmese teak forests.

Between 1853 and 1920, over 60,000 kilometres of railways were constructed to provide the transport network that enabled Indian production and markets to be fully incorporated into the worldwide colonial economy. The demand for railway sleepers, heavy construction timbers, building timbers and wood to fuel the locomotives was enormous. As the forests were logged further and further afield to meet the demand, it became apparent that the accessible forests might be completely logged out.

Dietrich Brandis (who became the first Inspector General of Forests) was brought from Burma to assess the extent of the problem, and recommended that an Indian Imperial Forest Service should be formed. This was duly set up in 1864 (Guha, 1983) and followed by the first Government Forest Act in 1865, which was not fully implemented. A new law was drafted and passed in 1878.

The debate surrounding this Act revealed the apparent contradiction between statement and policy which was a reflection of two powerful and divergent opinions – those of Dietrich Brandis and B.H. Baden-Powell, a colonial administrator and one of the architects of the Act. Brandis recommended the setting aside of areas of forest for village use and management, Baden-Powell, on the other hand, wanted power to be vested in the state for control over forests. In the 1878 Act, it was Baden-Powell's viewpoint that held sway and the state's role as custodian of all forest resources was confirmed (Buchy, in press; Gadgil and Guha, 1992b; Luthra, 1994) with large areas of forest reserved for national use.

In 1878, the stringent conditions of the colonial forest Bill were contested by the Poona Sarvanjanik Sabha, which pointed out that the maintenance of forest cover could more easily be brought about by

> taking the Indian villagers into the confidence of the Indian government. If the villagers be rewarded and commended for conserving their patches of forest lands, or for making plantations on the same, instead of ejecting them from the forest lands which they possess, or in which they are interested, emulation might be evoked between different villages. Thus more effective conservation and development of forests in India might be secured, and when the villagers have their own patches of forests to attend to government forests might not be molested. Thus the interests of the villagers as well as the government can be secured without causing any unnecessary irritation in the minds of the masses of the Indian population. (quoted in Gadgil and Guha, 1992b)

Despite the apparent conflicts within the colonial administration, the legislation was passed and the more liberal view of local people's roles, reflected in the statement above, was destined to be forgotten for another hundred years. The more comprehensive Forest Act of 1878 made formal provision for the demarcation of forests to meet the needs of the growing colonial economy. It was in this Act too that the first attempts were made to transform local people's rights into privileges – action that has continued to have profound consequences for Indian forest management.

To facilitate the implementation of the Act, the Forest Department was rapidly developed. Initially it was staffed with German foresters, while British students were sent to France and Germany for professional training. In 1885, a special training school was established in England to continue the training of foresters for India largely in French and German methods. Thus it was the scientifically advanced form of forestry that had been created to manage state forests on a large scale for the long-term production of industrial wood in France and Germany that provided the technical input to colonial forestry.

From rights to privileges

The desire to formalise the regulation of the forest sector led to the Indian Forest Department identifying the best quality forests and instituting elaborate administrative and legal procedures to effect their transfer to the state. The central question for the government was how to 'settle' the traditional practices of local people to obtain wood and other products from the forests. This was debated extensively in the columns of the *Indian Forester* during the 1890s and is reflected in strong statements made in 1891 which clearly describe the sovereign right of the state over local users' rights (as had already been asserted by Baden-Powell). It was recommended that colonial interests would be asserted most readily if it was:

It is the history of conflict over formalisation of rights that will also affect the degree of trust shown by local people to new joint forest management approaches which are also predicated on the formalisation of user rights.

There is scarcely a forest in the whole of the Presidency of Madras which is not within the limits of some village and there is not one in which, so far as the Board can ascertain, the state asserted any rights of property unless royalties in teak, sandalwood, cardomom and the like, can be considered as such, until very recently. All of them, without exception, are subject to tribal or communal rights which have existed from time immemorial and which are as difficult to define as they are necessary to the rural population . . . Nor can it be said that these rights are susceptible of compensation, for in innumerable cases, the right to fuel, manure and pasturage will be as much a necessity of life to unborn generations as it is to the present . . . (In Madras) the forests are, and always have been, common property, no restriction except that of taxes . . . was ever imposed on the people till the Forest Department was created, and such taxes no more indicate that the forests belong to the state than the collection of assessment shows that the private holdings . . . belong to it (Remarks by the Board of Revenue, 1871).

> . . . recognised that the customary licence of removal of forest produce, or of forest grazing was not a right of the native to a permanent easement inseparable from the estate but only a privilege enjoyed under the goodwill of the owner and until the land was required for purposes which rendered its continuance impossible or at any rate more or less prejudiced.

This period thus marked a significant change in local patterns of forest usage with the state asserting that local people's rights to use forests were to be replaced by privileges. Privilege and its application became the prerogative of the state to assign or to remove. Rights were rarely granted as the state could not exert the same degree of control over their extent and application. Thus in most areas of India rights were extinguished and replaced by privileges. This system required generally illiterate people to apply in writing for their rights to be recognised as privileges, if they did not wish to forfeit them. Although this principle applied to most areas of British India, its application was uneven and has led to some interesting and important anomalies today, including *nistar* rights recorded in several States (including Madhya Pradesh) where local people may have rights in forests distant from their settlements, and the *shamlat* lands of Jammu and Kashmir (Chatterji and Gulati, n.d.). The settlement of rights is discussed in detail below for Kullu and Mandi districts in Himachal Pradesh in northern India, and for Uttara Kannada in the Western Ghats of Karnataka State (see Sections 2.3 and 2.4). An understanding of the history of these settlements is of fundamental importance for the foresters of today, since it is these rights that are still practised that will determine the response of local people to joint forest management programmes.

Although the predominant view portrayed of the colonial government is that of revenue maximiser, it is interesting to see in the texts of the last century many of the same uncertainties and contradictions of today also being discussed then. Perhaps we have now come full circle, and should revisit some of the notions of the nineteenth century in order to further the understanding of how local people can be involved in forest management (Box 2.1).

In the end settlement practice varied across India from extremes where no local rights were recognised to some areas where privileges and rights were allowed. This regional variation was in part due to the varied strength of local resistance and in part to strategic interests in areas of the Himalayas (Guha, 1983). Provisions were made in the 1878 Forest Act to allow villages to have areas of third-class forest for their own use; these, however, were generally forests of low quality (often referred to as wasteland).

As can be seen, the path of policy development was not straight or without conflict, indeed there was great contention about the objectives of policy which continued into the 1890s and into the formulation of the 1894 Forest Policy. The policy debates were fuelled by various reports including an influential policy document arising from a report on the improvement of Indian agriculture,

Box 2.1 Conflicts and Contradictions in Colonial Forestry Policy

'The second class of state forests includes the great tracts from which our supply of the more valuable timbers – teak, sal, deodar and the like – is obtained. They are for the most part (though not always) forest tracts, and *encumbered* by very limited rights of user; and when this is the case, they should be managed mainly on commercial lines as valuable properties of, and sources of revenue to, the State. Even in these cases, however, customs of user will for the most part have sprung up on the margins of the forest; this user is often essential to the prosperity of the people who have enjoyed it; and the fact that its extent is limited in comparison with the area under forest renders it the more easy to continue it in full. The needs of communities dwelling on the margins of forest tracts consist mainly of small timber for building, wood for fuel, leaves for manure and for fodder, thorns for fencing, grass and grazing for their cattle and edible forest products for their own consumption. Every reasonable facility should be afforded to the people concerned for the full and easy satisfaction of these needs, if not free. . . then at low and not at competitive rates. *It should be distinctly understood that considerations of forest income are to be subordinated to that satisfaction.*'

'There is reason to believe that the area which is suitable to the growth of valuable timber has been over-estimated, and that some of the tracts which have been reserved for this purpose might have been managed with greater profit both to the public and to the State, if the efforts of the Forest Department had been directed to supplying the large demand of the agricultural and general population for small timber, rather than the limited demand of merchants for large timber. Even in tracts of which the conditions are suited to the growth of large timber it should be carefully considered in each case whether it would not be better, both in the interests of the people and of the revenue, to work them with the object of supplying the requirements of the general, and in particular of the agricultural, population.'

Source: IFS, 1894 (emphasis added)

written by Dr Voelcker in 1893. This report had a major impact on Indian forestry, where Voelcker's conclusions were that Indian forest policy should serve agricultural interests more directly. In reply to this observation in a 'Review of Forest Administration in British India' (IFS, 1894), the government stated that:

> The sole object with which State forests are to be administered is the public benefit. (A)nd the cardinal principle to be observed is that the rights and privileges of individuals must be limited, otherwise than for their benefit, only in such degree as is absolutely necessary to secure that advantage.

In its 1894 policy the government defined four broad categories of forest that served to secure three different objectives:

i) Environmental objective: Forests the preservation of which is essential on climatic or physical grounds

ii) Economic objective: Forests which afford a supply of valuable timbers

iii) Local needs objective: Minor forests and Pasture lands

Forest lands were allocated and demarcated according to these objectives, and rights settled or excluded depending on which objective was to be met. The debates of this period do indicate that there was some allowance made for local people. However, in many cases local people's needs were in reality subordinated to commercial interests, as becomes increasingly apparent by the increase in forest-related disturbances.

Agriculture versus forestry

The tensions between local usage and state control were mirrored in contentious relations between the departments of agriculture, revenue and forestry. For the Revenue Department land was most usefully deployed under agriculture from which taxes could be claimed. Much of the debate in the latter part of the nineteenth century concerned the degree to which forest land should be converted to meet the needs of a rapidly increasing population. At the time, one of the major arguments against the formation of a forest department was that it would incur greater costs than profits (Ribbentrop, 1900 quoted in Buchy, in press). The conversion of forest land to other land uses has remained a source of much controversy through to the present times, despite strong legislation which has attempted to restrict the alienation of forest land.

The Forest Department's position vis-à-vis the administration was not unchallenged as the Agriculture Department sought more land for cultivation and the Revenue Department sought to reduce expenditure and increase income through the conversion of forest land to taxable agricultural land. Just as in Nepal, in India in the early days of the forest service, forest officers were placed under the responsibility of the local revenue officer – the collector – and considered themselves to be inspectors and advisers with the executive role and associated fiscal responsibility assigned to the revenue officer. This position changed, however, with the advent of the new forest policy and led to the curtailment of the revenue officers' powers.

The inclusion of forest lands under the remit of the collectors had some consequences for the rate of deforestation. Since a collector's performance was measured by the extent of forest land relinquished to agriculture, as Buchy (in press) so clearly shows, the incentives to convert were great at a time when the income from one acre of cultivated land was as much as that from several acres of forest. Buchy provides an example of the revenue gained from forests over a 70-year period between 1871 to 1941. From one district in the Western Ghats (North Kanara) the revenue from

It should be remembered that under certain conditions the demands of agriculture are greater than those made by forest conservation . . . Demographic pressure on the land is one of the greatest difficulties which India must face . . . Thus, wherever there is a genuine demand for cultivable land which can only be satisfied by forest land, the latter will have to be abandoned without hesitation . . . and it must be very clear that nothing in the Act of 1878 limits the discretion of the local Governments which can, without previously referring to the central Government, divest forest land for the needs of agriculture, even should the latter have been classed as reserved

(GOI, 1894 quoted in Buchy (in press))

forests amounted to 35–54% of the total revenue of the Bombay Presidency. Over the same period, the annual returns from the Revenue Department varied greatly. In 1871, the amount was twice as high as that shown by the Forest Department, whereas in 1941, it was almost three times lower. These figures underline the changing role of forestry over this period and the realisation of its full economic potential as its products served the burgeoning needs of the state.

Despite the pressure to convert forest lands to agriculture, the Forest Department was successful in maintaining supplies to the railways and making large profits for the state. From its establishment in 1864, the Forest Department earned a surplus of revenue over expenditure that grew from 1 to 15 million rupees annually by 1914 (Schlich, 1922).

By 1914, 25 million hectares had been 'dedicated' as permanent state forests in British East India (now India, Pakistan, Bangladesh and Burma), and a further 40 million ha placed under the Forest Department's administration. Five hundred forest officers surveyed, mapped and planned the protection, utilisation and regeneration of these vast forests and some 15,000 Indian rangers and forest guards supervised the operations and prevented local villagers' encroachments (ibid.). In 1927 a second Indian Forest Act was passed which reiterated the major goals of the 1878 Act and its classifications of forest land. It is this Act which has provided the legal boundary for all forestry development from 1927 to the present day.

Although the policy and plans were primarily directed to supporting the overall colonial development strategy, they also contained provisions for protection, conservation, game-hunters and other interests. In response to the need to reafforest land, secure local labour and provide some incentive, **taungya** systems were introduced in Burma in the 1850s which allowed shifting cultivators to occupy an area of forest for 3–4 years on condition that they planted and tended the tree seedlings with their agricultural crops. The Forest Department was thus able to control the cultivators in their traditional practices and regenerate the forests with valuable species, usually teak.

Britain's loss of power after the First World War, and the realignment of its economy solely within the Empire, led to an immediate reappraisal of the importance of forests. No longer could Britain rely on exploiting an ever-expanding frontier; the Empire had to be made self-sufficient. This led to the relatively late establishment of the Forestry Commission in Britain in 1921, with a mandate to establish a flow of forest products to protect the economy against the vagaries of timber supply from other countries. Links were strengthened across the Empire between forest services, with the application of a uniform institutional model and policy, focused around a series of Empire Forestry Conferences held in Oxford. This, together with the establishment at Oxford of an Imperial Forestry Institute, whose function was to train foresters for export to the colonial territories, ensured the invulnerability of a model based on the development of 'a constructive forest policy whereby the sylvan resources of the Empire may be scientifically conserved and prudently exploited for the mutual benefit of the British Commonwealth of Nations' (Anon., 1922).

The demonstrated success of this model in India provided the impetus for its extension and replication to other territories administered by Britain. Following an inspection by a senior forester from India a forest department would be established under the same framework as in India. Between 1879 to 1901 departments were set up in Cyprus, Mauritius, Ceylon, Malaya and Nigeria, and before the First World War in New Zealand and Canada. The Australian states set up Forest Departments between 1877 and 1920. Generally the forests were operated under similar fiscal rules to those of India, where 20–30% of the revenue remained as surplus after costs (Troup, 1940).

Thus the forestry that was constructed during the colonial period had some clearly defined characteristics that were shared across the world (see Box 2.2). It was based technically on French and German forestry but adapted to local conditions. Its determining characteristic was the allocation of large tracts of forest to meet the objectives of the colonial state. The 'major' product was industrial timber, whereas the 'minor' products were the fuelwood, local construction materials, forest foods, and fodder obtained from the forests by small-scale producers and farmers.

Box 2.2 Characteristics of Colonial Forestry in India

- it was based technically on French and German forestry and the notions of scientific forestry
- it was managed by a corps of professional officers within the colonial administration
- it allocated large tracts of forest to production to support the colonial economy, and removed or reduced existing local uses
- contradictions in policy remained between the relative support to be given to local livelihoods and maximisation of revenue

2.3 Policy and Practice: The Case of Himachal Pradesh

Current forest use is a reflection of formal and informal rights of access, and the ability of forest users to assert their rights or force access where they have no rights. Rights and their exercise provide important evidence of local authority structures, and the ways in which individuals and groups manipulate relationships of power. This case study describes the evolution of forest rights in two districts of Himachal Pradesh – Kullu and Mandi – and illustrates the importance of understanding the historical context of forest usage before trying to intervene today in existing, if hidden, local systems of forest management.

Himachal Pradesh was constituted as a Chief Commissioner's

Province in 1948, with the merger of 26 princely hill states, among them Mandi, and 4 Punjab hill states, among them Kullu. This difference of historical background helps to account for minor legal and administrative differences of land-tenure classification which persist to the present. Further hill territory was transferred from the Punjab in 1966, and in 1971 Himachal Pradesh was accorded full Statehood within the Indian Union.

Settlement and exercise of rights in Kullu and Mandi

Contrary to indications from other areas of India (see Guha, 1989), the process of settlement of rights in Kullu and Mandi did not result in the termination of local people's rights, but rather their acceptance and formalisation. The process of forest settlement began in 1866 in Kullu and later followed the classifications laid down under the Indian Forest Act of 1878. Settlement in Mandi began much later and ended in 1917 with a Settlement Report written by H.L. Wright, a Forest Officer. It followed a similar pattern to the Kullu settlement, although the history of land tenure in Mandi was quite different from that of Kullu. Decisions about the proportion of land to be placed under the reserved or protected category were guided in part by the provisions of the 1878 Act. Debate surrounding the definition of these categories was fierce and is ably summed up by Mr Hope (Government of India Gazette, 30 March 1878, quoted in Anderson, 1886):

> There exists throughout India a vast mass of forests which are not reserves and for the most part never can be ... mostly because they have other purposes to fulfil and are needed for the current use of the people, grazing of their cattle, the thatching, repair and construction of their houses, and even (in some cases) the fertilization of their fields and the eking out of their slender meal, and to ensure with this view the provident and reasonable exercise of rights, the existence of which is not disputed, appears to be as essential a part of forest conservancy as the formation of reserves and the nursing of gigantic trees...

Kullu

The Punjab government decided that the bulk of forests should be placed under the protected category with very small areas constituted as reserved:

> The Kullu deodar forests are not on ridges far away from cultivation, but are in the immediate vicinity of villages. They are the daily resort of the people for the pasture of their cattle, for timber for their houses, for fuel, fodder, manure and agricultural implements. There would not be much difficulty in reserving 150 square miles of **rai** and **tos** forests (spruce and fir) in Kullu, which lie high up away from cultivation, but difficulties arise when the

Deodar forests in close proximity to village, Himachal Pradesh, India

> waste to be reserved is just what the people require for the supply of their daily wants. (ibid.)

These views were reinforced by other eminent forest officers, such as Ribbentrop (quoted in ibid.):

> What is required is the closure of larger areas of deodar-producing land, and there lies the difficulty, for these forests extend down into the permanent grazing-grounds, are mostly situated just above the villages, often honey-combed by cultivation, and yield the first spring crops of herbs and young branches, which are of great importance to the often very badly wintered herds and flocks.

This debate about the extent of land to be reserved continued, with some advocating limited rights for local people and others arguing for more extensive rights with a provision for their increase over time. Mr Lyall, the Revenue Officer, considered that:

> ...it would not be fair to the zamindars or their descendants to convert each man's free right of pasturage into a right to graze only a fixed number of cattle, which number could not expand, however much his family increased or his circumstances improved. (quoted in Anderson, 1886)

The argument between the Settlement Officer for Kullu, Alexander Anderson, and eminent members of the Forest Service continued and is recorded by Anderson in his discussion of the extent to which local people's rights should be extinguished in forests close to their villages. Anderson (1886) believed that it would be wrong to commute local people's rights:

Close association between forest and village, Himachal Pradesh, India

> It is scarcely necessary to touch on the proposal to commute the rights under Section 15 by cash payments. The people are dependent on these rights for their very existence, and the extinction of the rights would be most unjustifiable expropriation.

The outcome, in support of local people's rights rather than making them privileges to be rescinded at will, ensured that the area of land brought under reserved forests was small. This was accepted by the Punjab government in 1883:

> The Government of India has accepted the proposal of the Local Government that the bulk of the demarcated forests shall be treated as protected forests, such restricted areas as the Forest Settlement Officer and local Forest Officers may agree upon being treated as reserves on the understanding that this procedure will be carried out in an accommodating spirit. (quoted in Anderson, 1886)

Settlement of rights of local people in these different categories of forest in Kullu was concluded in 1886 by Anderson. The detailed settlement report still forms the basis for the current legitimisation of village-level rights. The decision to place most of the forest area under the protected category led to the final classification of forest land areas into four types based on a recommendation by Schlich (the then Inspector General of Forests).[2]

The differences between the categories of forest were based on their timber utility. Reserved forests were created in areas remote from habitation where there were limited or no rights, or in areas close to villages where there was sufficient other forested land available for use by local people. Good quality forests with a large number of rights were not reserved but placed in the protected category, thereby ensuring that local people were still able to exercise their rights. Class I forests were generally those remote from habitation containing valuable timber species such as deodar; rights were clearly defined in these forests. Class II forests were considered to be less valuable commercially and thus greater numbers of rights were permitted. Class II forests differed from undemarcated forests in that grazing rights were clearly defined and the land could not be alienated for cultivation. Undemarcated forests close to habitation were considered to be a land resource available for cultivation and the supply of grazing and tree product needs.

The jurisdiction of the Kullu foresters included a land area under the control of the Rai of Rupi, a local ruler. He was awarded control of the undemarcated 'waste', a ruling which has since led to the large-scale allocation of land by the Rai to those whom he favoured. This right was only rescinded in 1977, prior to which date villagers state that much of their undemarcated grazing land was allocated to private individuals either under the right of **nautor** or as favours.[3] The right of **nautor** gave farmers access to wasteland to expand their cultivation:

> The peasant proprietors of the **kothi** (an administrative unit) have a right to ask to be allowed to extend their cultivation in the waste of the kothi, and government has a right to refuse to permit it where it may seem necessary to refuse in the interest of forest conservancy, of the preservation of the hillsides from land slips or of the grazing rights of individuals. Otherwise, permission is given, and the peasant who breaks up the land becomes the proprietor without any payment of any price or of any rent charged other than a demand equivalent to land revenue. (Lyall, 1891, quoted in Singh, 1953)

Mandi State

At the time of forest settlement, Mandi was a princely hill state, a status which led to several important differences in forest land tenure between Kullu and Mandi. The current geographical area covered by Mandi district includes the former Mandi and Suket States; each had different forest settlements. However, the nature of the settlement is very similar. The rights as recorded are given in Box 2.3.

The different land-tenure structures arise out of different ownership structures. In Mandi, large areas of land were given as **jagirs** to members of the ruling family (a similar land-tenure structure was practised by the Ranas in Nepal). They employed tenant farmers to manage their land and exacted payment of half the crop. However, both tenant farmers and farmers who owned their land all shared the same rights of usufruct in forest lands. In Mandi these rights were known as **bartan**, and the right-holders as **bartandars**.

Under the terms of the settlement, **bartans** were only recorded in areas designated as protected forests, and not in undemarcated forests; exercise of rights in these areas 'will be governed by existing custom' (Emerson, 1917). The Settlement Officer considered the recording of **bartan** rights to be essential to protect the **bartandars** against those who had no rights illegally using the forest. This

Box 2.3 User Rights Admitted in Demarcated Protected Forests: Mandi and Suket States

- Grazing for cows and bullocks, buffaloes, sheep and goats, ponies and mules
- Firewood
- Grass cutting for fodder and thatching
- Timber for building purposes
- Timber for upkeep of temples
- Collecting of brushwood and thorny shrubs
- Collection of fallen leaves and needles
- Lopping of trees for fodder
- Collection of fruits, flowers, leaves, edible seeds, medicinal roots and honey
- Cutting of hill bamboo for baskets
- Collection of torch-wood
- Collection of resin and deodar oil for medicinal purposes
- Collection of wood for burning dead
- Collection of earth and stone for building purposes
- Cremation grounds
- **Thaches** (grazing areas) inside the forests
- Right of way and water
- Leaves and bark for tanning
- Charcoal for agricultural implements
- Trees and fuel for religious festivals and ceremonies
- Timber for **tans** (huts erected in the fields from which to watch the crops)
- Berberis for **dahay** (basket where hill bamboo is not available)

Source: Beotra, 1926

continues to be a concern commonly voiced by local people; without enforceable legal rights it is difficult to protect forest resources against those who have no usufruct rights.

The actual recording of **bartans** was carried out 'when the people of a whole **ilaga** or group of villages were present'. The chances of omissions or errors were thus greatly reduced, while the people were given full opportunity and encouraged to 'ventilate their views' (ibid.). It would appear that great care was taken to ensure that the rights recorded represented the actual use and needs of the users at that time (although a reading of the written report does not necessarily accurately reflect the actual implementation of the policy). **Bartandars** had responsibilities to protect the forest against degradation:

> ... the principle on which bartans are admitted is that the bartandars are responsible for the protection of the forest in which they enjoy their bartans. They are held responsible that no outsiders graze, that no trees are cut without permission, and that the forests are protected against fire. (Beotra, 1926)

The history of forest demarcation begins earlier than this first settlement with a demarcation carried out in 1889 by Maynard, Counsellor to the Raja. Two classes of forest were created: **siyan** or demarcated and **bartan** or undemarcated. The **siyan** forests were considered to be reserved and closed to rights 'not so much for the sake of forest conservancy, as for the provision of shooting preserves for the Raja, or as shelter belts for the old forts' (Wright, 1917). It was considered that this classification was inadequate because it took insufficient account of the needs of local people. Demarcation had to begin again. Thus it was a period of considerable uncertainty for local people where rights of access appear to have been removed and replaced at will. One can only speculate that this uncertainty may have led to increased deforestation.

The reasons for differences in forest areas now seen between Kullu and Mandi are clearly explained by Wright in his Settlement Report of 1917. Since the process of settlement started some 40 years later in Mandi, **nautor** allocations were virtually uncontrolled over this period:

> ... with the result that every ridge of moderate gradient has cultivation at intervals along its slopes, and in most nallahs (streams) all the warmer aspects have been taken for cultivation, while only the colder places have been left as forest.

As an accompaniment to this, more areas of agricultural land were in close association with forest areas, making it impossible to restrict rights in these areas:

> It will be realised, therefore, that such of the forests as adjoin cultivation are intimately connected with the life of the people

> and that any extensive scheme of closure is out of the question, for not only do the forests ... form the main grazing ground, but the people rely upon them for fodder and bedding for their cattle and manure for their fields. (ibid.)

Therefore more forest land was left undemarcated than in Kullu. However, the demarcation carried out by Wright attempted to redress this apparent imbalance by demarcating more areas of fir, spruce and pine forest; user rights were still permitted in these demarcated forests and followed the pattern of settlement carried out in Kullu.

In undemarcated forests all the rights allowed in demarcated forests were permitted in addition to the right of breaking land for cultivation. As in Kullu where a similar right persists, this has led to dissension and dispute between local people, and the Revenue and Forest Departments, as to whether breaking of such land is legal.

Trees on private land, according to the *Mandi-Suket Gazetteer* of 1904, belonged to the Raja. However, under the terms of these forest settlements it was decided that the land-holder should be given these trees for a nominal price. The reasons for this are interesting and again show a respect for local rights and a pragmatic approach to forest conservancy (perhaps somewhat at variance with experience from other parts of India):

> In the course of recording rights, it has been noticed that people claim such trees as their own, supporting their statement by the fact that the lands on which such trees are standing have already been assessed to their fullest capacity for product and further that the portion of the field over which trees throw their shadow and for which land revenue is realised does not yield any outturn. In the opinion of Mr Singh as the people have already in the past years tried to protect and reserve their trees against destruction and to allow the continuance of the same spirit which will eventually engender the feeling of respect for the protection of trees leading ultimately to the interest of Forest Conservancy ... only a lenient view in this connection should be taken. (Beotra, 1926)

Thus at an early stage, rights to trees on private and state land were clearly defined with the principal objectives of meeting state needs for timber and local needs for forest products.

Structure of forest protection

Local systems for the management and regulation of forest access existed prior to the British formalisation of management in both Kullu and Mandi. The **negis**, revenue collectors, had extensive powers over the distribution of forest products; for example they were allowed to allocate up to 40 pine trees for new construction or repair to existing buildings. These powers were considered to be too extensive by the settlement officers and were curtailed.

Rakhas, local forest guards, predate the imposition of British forest management and protection, but their position was recognised by the Forest Department and their knowledge of forests and local rights drawn on to revise the forest settlements for both Kullu and Mandi.

The **rakhas** were charged with drawing up lists (**jamabandi**) of **bartandars** which were then used in the settlement. According to Wright's report, such lists were in existence before British forest management was instituted. The **jamabandis** also detailed those who had grazing rights and the fees they should pay for particular types of forest rights. For example, miscellaneous fees were levied for the exercise of forest **bartans** and were collected by forest **negis** (locally appointed officials). The most common of these fees were **Banoli** and **Jungal ka Rakm**. They covered different types of use – for example, fees were charged in one area from people who grazed their buffaloes in a particular part of the forest; or for the preparation of agricultural tools; or in other places for the collection of firewood. Thus the settlements were formalising a system of fees for uses already in existence, and were not superimposing new alien systems.

The **rakhas** were retained by the British in their forest protection role until such time as the Forest Department had sufficient forest guards of its own. It was decreed that management of demarcated protected and undemarcated forests should be:

> ... through the agency of the negis (headmen) of the kothis who would be responsible for their proper management, while the Forest Department would merely control the action of the negis; the latter would grant trees to the people in accordance with the rules, assisted by the rakhas who are paid by the kothis. It is intended that the first class forests will be managed exclusively by the Forest Department without the assistance of the rakhas. This will probably not be possible in all cases without a larger establishment, and it may for a time be found necessary to utilise the services of the negis and the rakhas in the management of the more isolated and less valuable 1st class forests. (Anderson, 1886)

Forest rights in Kullu and Mandi

Although there were differences in the way the settlements were carried out in Kullu and Mandi, the rights admitted were similar in nature. Anderson's settlement allowed the following rights to be exercised without permission in all forests: to cut grass, to remove medicinal roots, fruits, flowers, dry fallen wood, except deodar (*Cedrus deodara*), walnut (*Juglans regia*), box (*Buxus* sp) and ash (*Fraxinus* sp), to cut bamboo and to take splinters of deodar and kail stumps (*Pinus wallichiana*). These rights were ascribed in full detail for each forest separately. The names of trees and shrubs that could be cut or lopped without permission were specified; the times during which manure leaves, whether dry or green, could be taken were fixed; paths through each forest were indicated in detail; the upland grazing areas or places where sheep were penned were named, and the times during which they were used were specified. The loppings of certain tree species had special conditions attached to them, and some could only be lopped to a certain height.

As all these rights were and are appendant to cultivated land, the right-holders are described not by individual names but by the name of the hamlet. Tenants were able to exercise the rights associated with the land they cultivated. Those without land or a tenancy did not have any rights in demarcated forest land but had to rely on undemarcated areas to satisfy their needs. Usage of these lands would continue only as long as other local people permitted.

The settlement of rights by a revenue officer as opposed to a forest officer appears to have moderated the effects of the 1878 Indian Forest Act. Anderson acted in a way to ensure that local people's rights were not unduly abrogated and were sufficiently flexible to allow for future change. This was in accordance with Government of India recommendations issued in 1883 (Anderson, 1886)

> ...it was laid down that the record of rights should show as accurately as possible the extent of the rights now existing in the forest and the condition under which and the localities in which they may be exercised; and that, while allowing a moderate increase or modification of the existing rights of the indigenous population, the record should prevent the indefinite growth of rights beyond what is now found to exist.

However, Anderson did identify several 'great' rights which were allowed to increase:

- the right to manure leaves, dry and green
- the right to building timber
- the right of grazing

Perhaps the most important and lucrative 'great' right still in operation today is the right of every land-holder to receive timber at reduced rates for the construction and repair of housing and other buildings. Land-holders are also permitted to remove free all pines, spruce and fir uprooted by snow or other causes. Timber thus acquired may not be sold to others and must be for the right-holder's own use (Aggarwal, 1949). The breakdown of the joint household system and the increasing population have led to increased demands for timber, and in some areas to the degradation of forests close to villages. Aggarwal in his working plan of 1949 suggested that it would be 'more equitable to insist on the local right holder paying a price for his trees which should, as far as

Use of timber from timber distribution rights to build large timber-framed house, Himachal Pradesh, India

possible, cover the cost of reproducing the timber which he consumes'.

Much recent discussion has centred on the curtailment of these rights, in particular for timber. However, any attempt by the Forest Department to regulate timber rights has led to political uproar and a rapid reversal to previous practices. Although, under the terms of the original settlement, provision was made to ensure that no rights were permitted to become injurious to the condition of the forest:

> If the existence of rights as admitted would endanger the existence of the forest a limitation must be placed on the exercise of those rights . . . (Anderson, 1886)

However, application of these rights has continued to be bound by rules laid down in the nineteenth century and has not been allowed to evolve with the changing economic, political and social environment.

Forests in the twentieth century

The period from the beginning of the twentieth century to the end of the Second World War was marked by massive exploitation of the forests to supply the infrastructural needs of India and the empire generally. Timber exploitation was carried out by contractors and regulated by the Forest Department. The two World Wars placed unsustainable demands for timber on the forests of Kullu and Mandi. During the Second World War the forests were overcut above and beyond the prescriptions of the working plans, leading to large areas of denudation which required later investment and regeneration.

In 1948, most of the princely states merged into what is now known as Himachal Pradesh, thus bringing all forests under one central controlling authority – the Chief Conservator of Forests. After 1948 more rigorous management was imposed and demarcation of forest boundaries was placed high on the agenda (Bhati, 1990). However, reports indicate a large degree of uncontrolled logging under contractors to meet the rapidly expanding demand for raw materials for pulp and building materials (Tucker, 1982). It would appear, at this time, that local right-holders were unable to prevent outsiders from over-exploiting their forest resources. The expansion into new areas of unexploited forest was helped by the massive road-building programme. Most forest areas of upper Kullu were inaccessible prior to 1950 when the first all-weather road was constructed up the Beas river gorge and into the Kullu Valley (ibid.).

The new era of modernisation and socialism placed different pressures on Forest Departments. In Kullu the right of **nautor** continued to be implemented by the revenue authorities, which led to large-scale allocation of Class III land to landless or marginal farmers. The increasing pressures on Class III lands which were supposed to act as a buffer between village needs and timber production forests were becoming unsustainable. Although these pressures were recognised by the Forest Department, agreements with villages to manage local forest resources better were not effective.

Modernisation and a changing political and economic climate also led to changes in the management and utilisation of forests. The Indo-China war of 1962 had a major impact on timber exploitation in Kullu district. As a result of the war many roads were constructed in previously inaccessible areas of Kullu, further opening up large areas of forest for exploitation (Tucker, 1982). Mechanised felling was introduced in Kullu leading to over-exploitation and massive regeneration difficulties.

Other changes in the local economy, notably the fruit and latterly the tourism industries, have also placed increasing and unsustainable demands on the forests both for raw materials and for expansion of cultivated land. The expansion of commercial apple production has been rapid: in 1948, barely 1,000 ha were under fruit trees; by 1970, the area had increased to over 44,000 ha and to nearly 150,000 ha in 1988. Apples constitute 80% of the fruit produced (Partap, 1991). The expansion of apple marketing also led to a large demand for packing cases, formerly manufactured from locally grown pine, and at higher altitudes, fir. The use of these species has now been prohibited and eucalyptus imported from the neighbouring States of Punjab and Haryana has largely replaced the use of local species for packing case manufacture.

Mosaic of orchards and small private woodlands, Himachal Pradesh, India

2.4 Experience from the Western Ghats, Karnataka

In contrast to the experience from Himachal Pradesh, events in the Western Ghats serve to highlight the differences that exist from one region of India to another.

In Kanara (now known as Uttara Kannara) in 1806, an officer was appointed as Conservator of Forests. The following year a proclamation was issued 'asserting the company's right of sovereignty over the forests, and forbidding the felling of timber by private individuals'. In the words of Brandis (quoted in Gadgil and Guha, 1992a) the object was to ensure:

> a regular supply of timber for public purposes from the public forests, the Conservator of Forests ... assumed much larger powers and apparently he was supported by the Government ... No attempt was made to settle the boundaries of these forests ... The Conservator of Forests extended his operations over the whole country; he cut down and appropriated to the use of the Government, not only the trees of the private forests, but even those growing on cultivated lands. The proprietor was compelled to pay duty on the timber growing upon his own property when he made use of it for his own purposes ... For the regeneration and improvement of all the forests the Conservator did nothing. It was complained of by all the local authorities ... In 1822, Sir Thomas Munro, then Governor of Madras, insisted on its being abolished, [observing] that 'no paltry profits in timber can compensate for the loss of their [people's] goodwill.

Uttara Kannara has a long history of sporadic uprisings against the British since 1800. The forest issue roused the majority of the people. One of the earliest expressions of organised protest against the forest administration was the convening of the Kanara Conference on Forest Grievances in 1884 in Sirsi. The 1878 Indian Forest Act had come into full force with the extension of reserved forest areas. Feeling grew against the forest restrictions over a 50-year period, and was managed by the British through a series of concessions. Some of the customary privileges were restored. Interestingly, the reason given for the lack of settlement of rights was that, since the forest areas were so extensive and the population density so low, it was considered unnecessary to formalise customary rights which were not a threat to forest conservancy.

Box 2.4 describes the actions of the Forest Grievances Committee set up to investigate the impact of the new forest policy, and indicates how this was the beginning of a long period of resistance to the new forest order which culminated in the famous forest **satyagraha** of 1930.

The process of settlements of rights and replacement with privileges was conducted by Forest Settlement Officers (FSO) who were often part of the Revenue Department and not the Forest Department. In one case cited, the Conservator of Forests for Kanara requested that the FSO should be 'an experienced officer, acquainted with the customs and needs of the local inhabitants as well as with those of the adjacent districts which depend on the forests of Kanara for their provision of timber and bamboo' (Government of Bombay, 1874, quoted in Buchy, in press). That such people were appointed is borne out in Himachal Pradesh, where, as we saw in Section 2.3, the settlement, carried out by Alex

Anderson a revenue officer, was extraordinarily careful in its identification and conservation of local people's rights. One can only assume that the same level of care was not applied in Uttara Kannara where low population densities and more extensive forests did not necessitate such detailed formalisation of rights, and also

Box 2.4 Vana Dukha Nivarana Sabha (1886) to the Forest Satyagraha (1930)

This committee (for the alleviation of difficulties related to the forest question at Sirsi) started work in 1886. It was widely supported by landowners and tenants alike and started a comprehensive inquiry into the replacement of rights by privileges. The inquiry focused on several important areas: access to the forest, gathering and use of minor forest produce for domestic needs; the status of artisans and their dependence on Minor Forest Products (MFPs), shifting cultivators and professional stockbreeders.

This was the first stage in resistance to the changes in the forest sector and is interesting in its use of the bureaucratic tools of committees, inquiries and the written word.

The Forest Satyagraha

In July 1930, at a meeting held in Hubli the Karnataka committee decided upon the following plan of action. The primary motivation of the participants was to recover rights of forest usage, as well as to obtain concessions for the extension of agriculture. The methods followed were:

- to begin as soon as possible work in the Sirsi, Sidappur, Kumta and Ankola **taluks**
- to violate forest laws every Monday
- to organise information centres
- to fell all trees in the Minor Forests, with the exception of *Terminalia*
- to systematically clear forests within a radius of 30 yards
- to exploit dead trees
- to fell sandalwood trees

In August 1930, three sandalwood trees, a few saplings and eight teak trees were cut at Sirsi. In other parts some 150 *Acacia catechu* were pruned and damaged. In September, a procession was organised to create unrest in 'special reserves', while others felled *Acacia* and teak trees elsewhere. The movement gained momentum in 1932–3 and the agitations became more persistent in the coastal areas, comprising up to a thousand people.

The peasants focused their attacks on the reserved species which had traditionally played an important role in local and household economies, but which also generated the greatest revenues for the foresters. They destroyed the investments of the Forest Department when they could, misappropriated sources of revenue and undertook to decimate the forest around their habitations.

Despite the evident participation of members of the Congress Party and the attempt to organise and supervise the agitation, the forest **satyagraha** emerged directly from the long period of unrest which had preceded the nationalist movement.

Source: Buchy (in press)

where national interests in high value forests would remain paramount.

The words of Voelker (1897), when he said that British forest policy in India had served to '...break up village communities so that by the end of the nineteenth century they became for the most part heterogeneous bodies rather than communities', should be recalled here when considering the differences between India and Nepal, and the applicability of approaches designed in different social and political contexts. In Nepal the penetration of the state into areas remote from Kathmandu was relatively poor; thus in many cases local management systems remained in place irrespective of policy change introduced by the government. Compare this with India where most regions were incorporated into the colonial economy and institutional framework, and where even by the end of the nineteenth century disintegration of local structures was already under way.

2.5 Age of Commercialism

The change in silvicultural policy at the end of the First World War heralded the start of commercialisation of the forest sector. From this period large areas of natural forest were replaced with uniform plantations of marketable species (for example teak in the Western Ghats). Thus the multiplicity of species that supported local people's livelihoods in a diversity of ways was replaced by monocultures of species valuable to the state. As Stebbing (1927, vol. IV) notes, this policy was completely rational for a government meeting its national economic goals:

> The natural forests, frequently under-stocked and composed of large numbers of species, many of which had little or no marketable value, were to be replaced by a comparatively small number of valuable species ... In 1927/28 it was calculated that plantations would have values seven to forty times those of the original forest.

It would be interesting to calculate whether there was a concomitant increase in value to local people with the replacement of the natural forest with plantations.

The two World Wars and the period of decolonisation were a time of rapid industrialisation with a matching demand for raw materials. In newly independent India, it was Nehru's policies of industrialisation rather than Gandhi's **gram swaraj** (village self-development) that were applied more vigorously. A National Forest Policy was declared in 1952 with the objectives of meeting:

> the sustained supply of timber and other forest products required for defence, communications and industry ... and the need for the realisation of the maximum annual revenue. (FRI, 1961)

By 1980, over 3 million ha of plantations were established, the major proportion of which were to fulfil industrial needs (CSE, 1982). Exploitation of forests was at its height, with the use of private and relatively uncontrolled contractors. Development of wood-product-related industries accelerated, with an emphasis on the indigenous conversion of forest products to end products. For example, in 1924 5,800 tonnes of bamboo were used for paper manufacture; by the beginning of the Second World War 58,000 tonnes were being used. This increased to over 5 million tonnes by 1987. From the mid-1970s, the government set up Forest Development Corporations whose main objective 'was to draw institutional finance into the areas of logging and harvesting, and plantation forestry' (Thakur, 1984). The strength of the Forest Departments also grew during this period, with an increase in staff numbers from 50,000 in 1961 to 93,500 by 1972 (Pathak, 1994).

The history of local resistance against forestry practices continued. Vigorous protest movements arose in many areas. In Bihar, in 1978, local people protested in what has been called the 'Tree War' against the replacement of natural forests by teak plantations (CSE, 1982). In the Himalayas, the Chipko movement protested against the logging of the pine forests, and in Madhya Pradesh protest managed to halt a World Bank project that was to turn 20,000 ha of natural forests that supported the economy of tribal groups, into pine plantations (ibid.; Dogra, 1985; Anderson and Huber, 1988).

Policies of village-level rural development in the Gandhian tradition were pursued concurrently but sometimes conflicting with industrialisation. For example, the Gandhian ideals of strengthening the position of small peasants and landless labourers through collective action found an expression from the mid-1940s in the formation of Forest Labour Co-operatives. These were set up to give forest workers, particularly forest-dwelling tribal groups, protection from exploitative relations imposed by logging contractors. The co-operatives assured a fair wage and a share in profits obtained from working the forests (Muranjan, 1974). However, after the formation of state forest development corporations the areas of operation of the co-operatives were transferred to these corporations. The notion of partnership between local people and industry disintegrated as the corporations considered that 'forest dwellers as the partners in the exploitation of the forests were not acceptable ... as [they are] organisations deliberately formed to generate profits for reinvestment into the area' (ibid.). The Forest Corporation in Maharashtra, for example, changed the principles under which the co-operatives had been working and decided to give the societies coupes on a logging-only basis without profit-sharing as had been the previous arrangement.

The further alienation of local people from forests was achieved through an influential report produced by the National Commission on Agriculture in 1976 (which incidentally also recommended the implementation of social forestry, as discussed below). This considered that 'the production of industrial wood has to be

'The reservation of vast tracts of forests, inevitable as it was, was . . . a very serious blow to the tribesman. He was forbidden to practise his traditional methods of [swidden] cultivation. He was ordered to remain in one village and not to wander from place to place. When he had cattle he was kept in a state of continual anxiety for fear they would stray over the boundary and render him liable to what were for him heavy fines. If he was a Forest Villager he became liable at any moment to be called to work for the Forest Department. If he lived elsewhere he was forced to obtain a licence for almost every kind of forest produce. At every turn the forest laws cut across his life, limiting, frustrating, destroying his self-confidence. During the year 1933–4 there were 27,000 offences registered in the Central Provinces and Berar. It is obvious that so great a number of offences would not occur unless the forest regulations run counter to the fundamental needs and sentiments of the tribesmen. A Forest Officer once said to me: 'Our laws are of such a kind that every villager breaks one forest law every day of his life'.

(Elwin, 1964 quoted in Gadgil and Guha, 1992a)

the *raison d'être* for the existence of forests' (GOI, 1976).

The experience in Himachal Pradesh where local people's rights were given priority over national interest provides an interesting counterfoil to experiences in other parts of India where the National Forest Policy of 1952 was more fully implemented. The policy clearly stated that the interests of the larger economy should prevail over local interests:

> The accident of a village being situated close to a forest does not prejudice the right of the country as a whole to receive the benefits of a national asset ... While, therefore the needs of the local population must be met to a reasonable extent, national interests should not be sacrificed because they are not directly discernible, nor should the rights and interests of future generations be subordinated to the improvidence of the present generation. (GOI, 1952)

This was further underlined by the National Commission on Agriculture (GOI, 1976) which reclassified forests as: protection forests, production forests and social forests (this mirrors the classification of the nineteenth century):

> Forests managed primarily for protection occupy hill slopes, watersheds of rivers, river banks, sea shores and other localities vulnerable to erosion and degradation ... Production forests which are commercial forests should comprise valuable or potentially valuable timber bearing stands occurring in favourable regions which are indispensable for development of the country and for meeting diverse requirements of the national economy... The social forests should cover waste lands, panchayat lands, village common lands on the side of roads, canal bank, and railway lines, which may be brought under forest plantations, shelter belts and mixed forestry comprising raising of grass and leaf fodder, fruit trees and fuel wood trees.

The major feature of the National Forest Policy of 1952 was to reinforce the right of the state to exclusive control over forest protection, production and management. At the same time, provision was made for the integration of the princely states into the Indian union, thus greatly enlarging the domain of the Forest Department. This led to several problems where inconsistencies arose between those areas administered directly by the British and others under princely rule. In Bengal, for example, large areas of forest were under private control and ownership. In 1945, the Bengal Private Forest Act provided the necessary powers for the state to gain control of the management of these forests (Stebbing, 1929). In essence, this could be considered a benevolent act of government, where concern for the health of the forests prompted the desire to bring them back under 'systematic management'. However, commentators have pointed out that many of these so-called 'private' forests, although owned by one household, were shared by many villagers surrounding the forests, where usufructs were freely allowed. Under state control such free extraction of usufructs was not permitted. Many older villagers in these areas of West Bengal, and also in the Chotanagpur plateau of Bihar, mark the start of forest degradation from this time of nationalisation (Malhotra and Deb, 1991). They say that the landowners were not prepared to let the government gain the benefits from the forests and ownership of the land, and instead deforested large areas of these lands and registered them as private wastes or agricultural land.

As more forest lands were brought under state control, the demands of the commercial-industrial sector began to replace the strategic colonial needs (see Table 2.1). This was given a further push under the industrialisation policies of the 1950s. As industrial demands began to outstrip the sustained yield productivity of the natural forests, this led to calls for plantations of fast-growing uniform industrial timber and the replacement of natural forests with such biomass. In the 1960s, financial incentives were provided by central government to encourage the States to take up industrial plantations. In a far-ranging series of studies, Gadgil *et al.* (1983) show that the plantation programme, although in some cases successful in the production of industrial biomass, had wholly negative social consequences in those areas where local livelihoods were dependent on products from natural forests. Where there have been large-scale plantation programmes there have also been local protests. In Uttara Kannara in the Western Ghats of Karnataka, the plantation of teak was resisted by local people and protest continued against the policy of planting natural forest areas with eucalyptus.

Table 2.1 Four stages of industrial forestry

Period	Method	Species	Agency	Prime Beneficiary
1947	Selection felling	Indigenous commercial species	Forest Department	Industry
1960–85	Clearfelling and monocultural plantations	Chiefly exotics	Forest Department	Industry
1975	Farm forestry	Chiefly exotics	Commercial farmers	Commercial farmers and industry
1985	Imports and captive plantations	Exotics	Joint sector	Industry, importers

Source: Gadgil and Guha, 1992b

Eucalyptus plantation, Bihar, India

The application of industrial forestry practices changed to reflect different client groups and objectives; with this came different management and silvicultural interventions which in their turn had differential impacts on the livelihoods of forest-dependent groups. As can be seen in Table 2.1, the period of maximum impact occurred during the 1960s-mid-1980s with the conversion of natural forests to plantations and the associated degradation in species diversity and livelihoods dependent on this diversity (see Gadgil *et al.*, 1983 for a discussion of the impact of the Bastar Pine Plantation project in Madhya Pradesh which aimed to replace 40,000 hectares of natural forest with tropical pine to supply raw material for industry).

This same period marked the emergence of a vociferous and influential environmental movement associated with grassroots action. It was the beginning of an important relationship between local people, intellectual activists and the growing international environmental movement. This is perhaps best illustrated by the Chipko movement which gained enormous international attention in the 1980s (Bahaguna, 1984; Tucker, 1984; Shiva *et al.*, 1985; Guha, 1989; Weber, 1988; Mukul, 1993; Aryal, 1994; Pathak, 1994).

At the same time as local people's voices were being heard through protest against forest policy and practice, the advent of new social forestry practices was again to be the source of controversy and conflict.

2.6 Age of Conservation

The National Commission on Agriculture provided the following rationale for social forestry:
'Free supply of forest produce to the rural population and their rights and privileges have brought destruction of the forests and so it is necessary to reverse the process. The rural people have not contributed much towards the maintenance or regeneration of the forests. Having over-exploited the reserves they cannot in all fairness expect that somebody else will take the trouble of providing them with forest produce free of charge. One of the principal objectives of social forestry is to make it possible to meet these needs in full from readily accessible areas and thereby lighten the burden on production forestry.'
Compare this statement with that made in 1878 by the Poona Sarvanjanik Sabha, p.28

There are two interlocking threads running through this period: social forestry to meet local needs from non-government forest lands, and conservation forestry on government forest lands to preserve what was left of the rapidly degrading forest ecosystems. This latter thread was pursued with some vigour by Indira Gandhi, the then Prime Minister, who made many statements about the importance of conservation and left a legacy of highly protection-oriented conservation legislation.

Social forestry 1970s–80s

The most significant change to occur in the forestry sector happened in the mid-1970s with the report of the National Commission on Agriculture. This was a report that heralded the beginning of social forestry and an admission that the needs of local people for forest products were not being met, although, as has been shown above, it was tempered by the overall requirement that forests should meet the needs of the nation first and also be 'project-oriented and commercially feasible from the point of view of cost and return' (GOI, 1976).

Social forestry also brought a whole new series of actors on to the Indian forestry stage in the shape of international donors. With the increasing donor interest in support for the forest sector to supply fuelwood and other basic needs, social forestry seemed to fulfil the necessary criteria. Over a 15-year period, US$ 400 million were spent on establishing social forestry programmes (Poffenberger, 1990), and in a five-year period between 1979 and 1984 it was estimated that over 2.5 million hectares of land had been reforested (Guhathakurta, 1984). No longer was forest policy to be determined solely by national priorities, it was also now heavily influenced by the requirements of the international community. In some respects, Indian forestry could be considered to have become 'donor forestry' (Dargavel *et al.*, 1985; Chambers *et al.*, 1989).

Social forestry, in India, has been variously defined (as indicated in Chapter 1). In essence, it has the stated objectives of social change, increasing access to resources for the poor and landless, alleviating poverty and contributing to the development of villages (Fernandes *et al.*, 1984; Tewari, 1984) (see Box 2.5 for a description of the various perceptions of social forestry). Some of its main attributes were to provide alternative sources of fuelwood outside Forest Department-controlled lands. It took the form of farm forestry in block plantations and on field margins, and monoculture plantations on common and other government lands such as roadsides and canal and railway banks.

With this diversity of perceptions and extravagance of objectives, it is not wholly surprising that social forestry failed to achieve many of its desired consequences. However, what it did achieve was a renewed recognition of the forest-dependent nature of many rural people's livelihoods, and it paved the way for new forms of partnership to be developed. Some critics of social forestry noted

Box 2.5 How Many Faces Are There to Social Forestry?

- 'The term 'social forestry'. . . is deliberately avoided. This is because many people now feel it to be misleading since it implies that any form of tree growing by farmers or local communities automatically brings social benefits' (Foley and Barnard, 1984)
- 'Social forestry is the establishment of wood-forage-food production systems on uncultivated lands . . . It is to reduce destructive pressures on forest resources by providing economic alternatives to villagers who presently depend on forest exploitation for their livelihoods. It is to improve the lot of these villagers by intensifying production on uncultivated, unforested lands'. (Romm, 1981)
- 'The objectives of social forestry as defined by the National Commission on Agriculture . . . 1) supply of fodder, 2) supply of small timber, 3) supply of fuelwood to replace cowdung, 4) protection of agricultural fields against wind and soil erosion,and 5) creation of recreational amenities. Its main components are 1) farm forestry, 2) rural forestry and 3) urban forestry' (Pant, 1980)
- Social forestry projects are 'meant to bring a social change, to ameliorate distortions in the economy and to ensure a more equitable distribution of income and more equitable distribution of decision-making powers' (Srivastava and Pant, 1979)
- '[the purpose of social forestry] is the creation of forests for the benefit of the community through active involvement and the participation of the community. In the process, the rural environment will improve, rural migration will reduce, rural unemployment substantially cease . . . the overall concept of social forestry aims at making the villages self-sufficient and self-reliant in regard to their forest material needs'. (Government of Karnataka quoted in Fernandes and Kulkarni, 1983)

Source: Dargavel *et al.*, 1985

that one of the major problems was that what was required was a 'social forest policy rather than social forestry' (Kulkarni, 1983). This mirrors the question posed at the beginning of this Study Guide, where we questioned whether participatory forestry is a fundamental change in forest policy or practice, or is just the old forestry dressed up in new words (Pardo, 1985).

> Community forestry cannot be imposed from above and carried out in the face of hostile population. New forms of land-use impinge upon, and are influenced by, the daily activities of everyone. When the local people are not active participants and supporters of a project, saplings have a way of disappearing overnight. With fodder usually as scarce as firewood, uncontrolled goats or cattle can quickly ruin a new plantation even when disgruntled peasants facing the alternative of a lengthy hike to collect fuel do not covertly cut the saplings themselves . . . Community involvement, then, is not just an ideologically appealing goal; it is a practical necessity if rural forest needs are to be met.
>
> (Eckholm, 1979)

The formation of the National Wastelands Development Board (NWDB) in 1985 was one indication of the importance attached by government to the apparent problems of forest product supply for local people (GOI, 1990; Chowdhury, 1992). This shift of focus from the Ministry of Environment and Forests to a new Board was significant, as it heralded the removal of control from the foresters and the beginning of a reduced influence by professional foresters on policy-making (Chambers *et al.*, 1989). It also marked the beginning of an increased 'projectisation' of funds where foresters were expected to carry out activities in the context of projects rather than planning holistically for the total management of forest resources.

The initial remit of the NWDB was to reforest the so-called 'wastelands' of India. Its aim, as described by Rajiv Gandhi, was to afforest an ambitious 5 million hectares every year as fuelwood and fodder plantations (according to Chambers *et al.*, 1989 this is equivalent to an extraordinary 10 billion trees, or about 17,000 trees per village per annum). Just as the ideas of village forests and then social forests had been currency for 100 years, so too was the notion of 'waste'. It had been ably used by the colonial authorities to assert sovereign rights over areas of land that fell outside the purview of conventional land management. Thus Baden-Powell (1874) was able to state that: 'There never has been any doubt that in theory, the 'waste' – that is, land not occupied by any owner or allotted to anyone – was at the disposal of the ruler to do what he liked with; in short, was the property of the State'. In this way large areas of land used by local people for grazing, collection of medicinal plants, etc. were alienated and placed at the disposal of the state to allot as it deemed appropriate (see Box 2.6). Throughout the last century much emphasis has been placed on conversion of the 'waste' to more productive use, generally meaning its afforestation with commercially significant trees. This general trend changed in the 1980s when wastelands were again identified as a target area for intervention but this time as the land on which to grow the nation's fuelwood supplies. However, this still ignored the existing user rights to these lands.

The inevitable consequence of Rajiv Gandhi's target was the misappropriation of land that was under other forms of management – in particular, grazing – leading to the disenfranchisement of a large group of villagers, and a trail of failed plantations (Jodha, 1995).

In most areas where there were externally funded projects, plantation targets were rarely met. For example, in Uttar Pradesh a target of 3,080 hectares of woodlots was set but by the end of the project only 136 ha had been established (Cernea, 1992). The reasons indicated for this dismal achievement are provided by the World Bank:

> Poor villagers in Uttar Pradesh proved unwilling to contribute their labour as expected by the project in exchange for rather limited potential benefits from a small woodlot, after many years of protection and maintenance ... The social forestry organisation lacked relevant know-how and resources to deal with the socio-logical and technical problems associated with densely cultivated areas and very small farms (World Bank, 1985 quoted in Cernea, 1992).

Social forestry was used only as an adjective to turn on the channels of money. The officials and contractors ran the show. Once the channels of money dried up, the plantations disintegrated.
(Unnikrishnan, 1994)

Critical articles from a series of activists gathered evidence to indicate that social forestry was being used as a coercive tool for reforestation and in many cases was leading to greater local difficulties (Chambers *et al.*, 1989). Calls were made for the voices of local people to be heard, and increasingly local action was used to oppose unwelcome development interventions (see Pathak, 1994 for a detailed description of some of the action taken against social forestry activities and Box 2.7).

What is instructive from a brief reflection on the lessons learned from social forestry is the degree to which rhetoric and practice can diverge when institutional structures to support the achievement of objectives have not first been developed (Baidya, 1985; Roychowdhury, 1994).

'Social forestry programmes designed to provide fuelwood and fodder for the poor are instead becoming a source of quick money for big farmers. The Uttar Pradesh Government's World Bank-assisted social forestry programme has overshot its farm forestry targets by 3,430%, but fallen short of its targets for creation of community self help woodlots by 92 per cent. Judging by the World Bank's own mid-term review of the social forestry projects in UP and Gujarat, big farmers and the paper mills they supply with wood for pulp are emerging as the primary beneficiaries of these multi-care schemes.'
(*Indian Express* quoted in Dogra, 1985)

Box 2.6 'Waste' as a Gift

'Up to a very few years ago the district officer had sole control of all lands in his district and the unoccupied lands were his chief means of conferring patronage; he could give or lease them, or confer or confirm privileges in them. If troubled with lawless tribes of **badmarshes** he could offer them land as an inducement to settle to honest pursuits, and if a keen **shikari** (hunter), the forests were his sole and undisputed game preserves. Apart too from all departmental and personal interests his standpoint is different, his recollection carries him back to days when the forest as such yielded so little revenue that it was often as well to let the people help themselves to its products and graze their cattle in it, as to be at the worry and cost of collecting the revenue: when every acre broken up for cultivation yielded more revenue than a hundred acres of forest land besides enlarging the capabilities of the district, and promoting the well-being of the people; to a time when in fact the amount of forest broken up for cultivation became the recognised measure of a district officer's capability and tact' (Amery, 1876)

Source: quoted in Guha, 1983

The environment and the NGO movement

During this same period of social forestry, the growing focus on rural development and the plight of the rural poor led to a growth in the non-governmental organisation movement (Pathak, 1994; Gadgil and Guha, 1992c). The emergence in the 1970s and 1980s of

Box 2.7 Criticisms of the Social Forestry Era

- the peasantry was assumed to be a non-stratified homogeneous group represented through the panchayat, and thus access to benefits would be equally distributed
- local participation was limited to discussions between senior panchayat officers and the Forest Department
- it was assumed that the panchayat would represent the interests of its diverse constituencies
- planting of common lands replaced other existing uses of the land and led to local losses in livelihoods
- costs of protection (borne by Forest Departments and projects) were too high and unsustainable
- survival rates of trees were very low as plantations were considered to belong to the government rather than to local people
- land brought under social forestry schemes was reclassified as protected forests, and thus it became a forest offence for local people to collect products from the plantation areas
- fast-growing species were preferred by Forest Departments because of their ease of production. Although the high market value was of interest to certain local groups, many of those previously using the plantation areas were interested in access to non-commercial biomass
- access to intermediate products such as twigs and grasses was often denied to local people
- the very people, social forestry was supposed to benefit – the poor – were demonstrated to have gained little or nothing from the programme

Sources: Alvares, 1982; Shiva *et al.*, 1982; Shiva and Bandhyopadhyay, 1983; Mahiti Team, 1983; Sen and Das, 1987; Arnold and Bergman, 1988; Arnold and Stewart, 1991; Poffenberger and Singh, 1992; Saxena, 1992; Kaul, 1993 and derived from Pathak, 1994

the Chipko movement and the growing prominence of activists such as Anil Agarwal brought a new impetus and balance to policy dialogue. The Centre for Science and Environment in Delhi has been highly influential and much of the pressure to establish the National Wastelands Development Board came as a response to the increasingly strident calls from the NGOs to meet the fuelwood needs of local people. As concern grew during the 1970s about the rapid destruction of forests, Indira Gandhi asserted the centre's control over the States by transferring forestry from the State List to the Concurrent List by the 42nd Amendment of the Constitution in 1976. This led to a considerable strengthening of central control over forest lands which was further enhanced in 1980.

In that year, some of the NGO energy was focused in opposition to the new Forest Conservation Act which prohibited State governments from assigning by lease or other mechanism any forest land without the previous sanction of the central government. The Act was further strengthened by an amendment in 1988 which restricted the planting of horticultural crops and medicinal plants on forest lands. This provision has far-reaching consequences for the development of new management approaches to forest land under joint forest management and remains a major disincentive to local initiative and support for local livelihoods (Chambers *et al.*, 1989).

Under the Act much of the previous practice of allotting forest land to agriculture was now prohibited. In the main, the Act was intended to prevent the large-scale allocation of forest lands to major infrastructural projects. In reality, however, these allocations continued and it was the small-scale conversions that were prohibited (Pathak, 1994). This followed a stringent Wildlife Protection Act in 1972, which created a vast network of strictly protected parks and sanctuaries. In a further amendment to this Act, local people were totally excluded from these areas. Inevitably these areas of maximum importance for conservation are often areas where there are forest-dependent peoples (see Raval, 1994 for an interesting discussion of these issues with respect to Gir National Park).

In 1981, a draft Bill emerged for a Forest Act to replace the still extant 1927 Act. The Bill produced an enormous outcry from a wide range of organisations, with some saying that the Act should be renamed the 'Indian State Monopoly over Forest Act and Indian Forest Offence Code' (Anon. 1981). The draft Bill provided the focus necessary to galvanise diverse groups into action and in many ways acted as the catalyst to bring to the forefront the views of the environmentalists and social activists who had long claimed that government actions in the forests were not leading to the improvement of local livelihoods and were more in support of the needs of industry and capital. The Bill rather than requiring the involvement of local people in the management of forests, instead makes many references to the penalties to be imposed against them for infringement of forest rules (D'Abreo, 1982; Guha, 1983; Kulkarni, 1982 and 1983). In a period when social forestry was much vaunted as the new approach for solving India's forest problems, there was no mention of it as a possible mechanism for increasing the role of local people in management.

Interestingly, as will be discussed below, many of the criticisms levelled at the 1981 draft Bill are the same as those being used today against a new draft Bill brought out in 1994 (Roychowdhury, 1995). These criticisms include the comment that, just as there was no mention of social forestry, there is no mention of joint forest management approaches in the new bill. Does history have to continue to repeat itself like this, or is it the case that arguments have progressed, and a new relevant legislative framework will be introduced? Chapter 8 considers these questions in more detail.

Given the level of opposition to the draft Bill of 1981, it is perhaps not surprising that it was not formalised as the new forest act, and practice in India continued to be guided by legislation written by a colonial authority under a very different social, economic and political regime. Despite all the contradictions inherent within the policy and legislative frameworks, changes in practice continued to occur, with Forest Department staff and NGOs experimenting with more participatory approaches. However, it did not seem possible that there could be any recognition of local people's user rights in forest lands owned by the state (Commander, 1986).

2.7 Age of Collaboration

This new era dawned with the 1988 National Forest Policy with its explicit emphasis on participation of local people in the management and protection of forests. In Orissa and West Bengal two Government Orders were passed, in 1988 and 1989 respectively, which laid the foundations for the introduction of a central government resolution for joint forest management Kant *et al.*, 1991). Two years later, in 1990, a watershed in Indian forest history was reached with the passing of the Joint Forest Management Resolution (Box 2.8).

This has paved the way for 16 States to pass their own JFM resolutions which allow formal joint partnerships to be developed between local people and Forest Departments for the management of forests. This approach was supported in the Eighth Plan where the deficiencies of social forestry were noted and calls were made for a more participatory approach to be developed focused on the formation of partnerships between local people and Forest Department staff. Thus it was not until 1990 that joint forest management entered the vocabulary of rhetoric. However, implementation of such approaches long pre-dates this time, and indeed the resolution was based on the experience gained in West Bengal, Haryana and Gujarat over the previous 15 years (see Box 2.9 and SPWD, 1992 for description of the developments in West Bengal; for Haryana see SPWD, 1984; Singh *et al.*, 1984; Chambers *et al.*, 1989; Dhar *et al.*, n.d.; Scott and Gupta, 1990; Arnold and Stewart, 1991; Singh, 1992; Varalakshmi *et al.*, 1993; Kaul and Dhar, 1994; for Gujarat see Pathan *et al.*, 1991).

The origins of joint forest management lie in two experiments undertaken in West Bengal and Haryana. Here, where forests had

Box 2.8 Principles of Joint Forest Management

'The National Forest Policy, 1988, envisages people's involvement in the development and protection of forests. The requirements of fuelwood, fodder and small timber such as house-building material, of the tribals and other villagers living in and near the forests, are to be treated as first charge on forest produce. The policy document envisages it as one of the essentials of forest management that the forest communities should be motivated to identify themselves with the development and protection of forests from which they derive benefits'. (GOI, 1990 quoted in SPWD,1993)

- it encourages the development of partnerships between local people and Forest Departments to manage these forest lands jointly
- it provides legalised access for the local communities to adjacent forest lands
- it encourages local people to protect forest areas, to prevent free grazing of livestock and to assist in preventing illegal activities by outsiders
- it assures local people of a certain proportion of the intermediate and final harvests from the forest lands protected by them

Source: Arora and Khare, 1994

Box 2.9 From Conflict to Cooperation

'In the early 1970s, I was Divisional Forest Officer of Purulia Division, situated in the south-west corner of West Bengal. As a young DFO I organised raids with great zeal to recover stolen forest produce from all over the district. During one such raid, in June 1973, we encountered stiff resistance from the people bringing in the produce, which led to the police opening fire. Two people were killed and three injured. A number of forest and police personnel were also injured by missiles hurled by the miscreants. This incident resulted in my telegraphic transfer from the district, a judicial enquiry into the firing incident, and a government order discontinuing hat (market) raids. The staff were totally demoralised and the illegal trade continued to flourish. This was in fact a turning point in my career as I became convinced that there was no alternative to joint forest management if forests were to survive.'

Source: Palit, 1993a

degraded badly, Forest Department officials and local people began to work in partnership. Forest protection committees (FPCs) were formed, with the earliest starting in Arabari in West Bengal in 1972. In each case these committees were given the responsibility of protecting degraded forest land from illegal cutting, fires, overgrazing, and encroachment, and in return were granted access to a range of non-timber forest products. In the Arabari case the State government sanctioned the sharing of the coppice pole wood harvest in the regenerated **sal** (*Shorea robusta*) forests, giving 25% of the net returns to the village protection committees involved (Poffenberger, 1990; Palit, 1991: Pathan *et al.*, 1991; Dhar *et al.*, 1991). In Haryana, following the successful implementation in Sukhomajri village of a locally managed integrated resource management approach, autonomous Hill Resource Management Societies (HRMS) were also established in neighbouring areas. The focus of the intervention was not directly on forests but on the harvesting of rainwater through the construction of earth dams in the forests, the water from which was used to irrigate the cultivable land of small and marginal farmers. A fourfold, almost immediate increase in agricultural production provided the villagers with an incentive to protect the dam catchment from grazing so as to prevent the siltation of its storage capacity. Not all villagers could be expected to share the costs of protection equitably (by forgoing grazing rights) without the assurance of equitable access to the benefits of protection (through increased agricultural production). Therefore, all households were allocated an equal share of the water, irrespective of land ownership or size of holdings, in return for not grazing in the hills. Those without land or with very small holdings could use their share of the water on land rented from larger landholders, thereby gaining access to a share of the increased agricultural production (Sarin, 1995).

To date about 40 autonomous Hill Resource Management Societies are protecting approximately 1500 ha of forest land under joint forest management (Sharma, 1994). Unlike the long gestation of benefit-sharing from poles or timber in West Bengal and other States, Hill Resource Management Societies become the Forest Department's effective partners in forest resource management almost immediately by providing priority access to annual fodder and **bhabbar** grass (*Eulialopsis binata*) leases at concessional rates (ibid.).

Box 2.10 indicates the current status of joint forest management arrangements.

'The territorial aspirations of foresters are accompanied by claims to monopoly over scientific expertise; the aesthetic longings of nature lovers are legitimised by talk of biological diversity and the ethical responsibility of humans towards other species; the profit motive of capital masquerades as a philosophy of progress and development; and the requirements of tribal and peasant communities are cloaked by an ideology that, in a manner of speaking, opposes country to city and Bharat to India.'

(Guha, 1994)

2.8 Discussion

Although the State Forest Departments together control over one-fifth of the land area of India, and thus are the major players in any shift in policy and practice, the history of forest management, as has been seen, is not single-faceted but is characterised by conflict and debate over roles and responsibilities. The diversity of interest groups and stakeholders involved in influencing the direction of forest policy and legislation is great. There are, however, four identifiable groups that have had a major impact over the last 100 years: conservationists; foresters; industrialists; and the social activists (Guha, 1994). Policy is in a sense an amalgam of these voices, in some instances one group's voice gaining pre-eminence over another. Table 2.2 compares the three policy statements issued over the last 100 years, and it is interesting to trace how different voices predominate in each of these statements. What is apparent is that the 1952 policy was the most stringent in its opposition to local people's use of forests, and the provision for local people's use of forests in the 1894 policy is surprisingly generous (Chopra, 1995; Palit, 1993a).

If the same comparative table were to be drawn up for the three Forest Acts (1878, 1927 and the draft Bill of 1994), unfortunately the same progressive policy changes would not be reflected in the legislative framework. The 1994 draft Bill has raised many questions about the commitment to new management approaches, as

Box 2.10 Current Status of Joint Forest Management

- 16 State governments have issued JFM resolutions
- Several resolutions have been amended in the light of experience
- These 16 States have 74.6% of the country's 75 million ha of public forest land and 91.4% of the country's tribal population
- By mid-1992 more than 1.5 million ha of forest land were being protected (mainly through JFM arrangements) by more than 10,000 community institutions in 10 states
- There are indications that many forests are recovering with remarkable vigour and diversity
- Local community institutions are protecting their forests more effectively than State Forest Departments

Source: adapted from Sarin, 1995

has the proposed introduction of lease forestry for the allocation of wastelands to industry (Chopra, 1995).

In the case of the 1994 Bill, for example, it could be considered that the conservationists' voices have been heard above all others. Ultimately, it is not clear which voice will predominate or perhaps it will be an amalgamation of all four, although recent commentaries have indicated that the NGO social activist voice is not so strongly heard as it was in the early 1980s (Sharma, 1995). What is clear, however, is that the major stakeholders – the local people – do not have a direct voice in policy formulation (Dasgupta, 1995).

Table 2.2 Policy statement comparison 1894–1988

Major goals of 1894 policy	1952 policy	1988 revision
Access and rights • to restrict and regulate access of neighbouring villagers to the forests	• national interests considered to be paramount, above those of local people	• forest products as a right to be reserved for use of neighbouring communities
Conservation • to maintain hill slopes as Protected Forests in order to preserve climatic and physical condition	• continued conservation emphasis with stipulation that 60% of land in mountainous areas to be maintained under forests	• no change
Revenue • to derive revenue for the state from exploitation of valuable timbers	• objective modified with emphasis placed on increasing productivity	• direct economic benefit subordinated to the principal aim of ensuring environmental stability and ecological balance
Classifications • to allow access of villagers to 'inferior' forests under regulations to protect them from their own improvidence	• functional classifications based on end-use: 1) protection forests for physical and climatic considerations; 2) national forests for serving defence, communications, industry and other purposes; 3) village forests for meeting community needs; 4) tree lands, areas outside forest management control for amelioration of physical conditions of the country	• emphasis on serving interests of local communities including those living within and around forests; ignored legal and functional classifications of 1952 policy
Cultivation • despite restrictions on access, forest lands could be released for permanent cultivation	• required forests to be a recognised land use; called for land-use policy based on land capability and maintenance of at least one-third of land under forests	• further strengthened case for forests as a land use by re-emphasising the law passed against diversion of forest lands to non-forest use except with approval of central government

Source: Femconsult, 1995d

Reflecting on the recent changes in forest management practice, we are forced to return to a consideration of whether joint forest management is social forestry with a new name? As can be seen in Table 2.3, there are certain fundamental differences, in particular the location of activities on government forest lands rather than on revenue and panchayat lands, and the emphasis on partnership. However, the questions we shall be exploring in Chapters 4 and 5 look in more detail at the implementation of approaches, and it is

Table 2.3 What is new about joint forest management?

Social Forestry	Joint Forest Management
1. Objective: • satisfy local needs through fuelwood plantations to divert pressure from natural forest. Mechanism to be used – 'People's participation'	**1. Objective:** • meet local needs equitably for diverse range of forest products through natural forest regeneration under community protection • community empowerment to make decisions with Forest Department as joint partner
2. Who • private farmers (especially larger farmers with credit access) • 'communities' through the panchayat system, but without identifying particular social units	**2. Who:** • clearly defined and organised community user groups (formal/informal) supported by the Forest Department • focus on the most forest-dependent-women, tribals and landless
3. Where: • private lands • common property (revenue lands, village grazing, panchayat land, ill-defined tenure)	**3. Where:** • state forest lands (protected and reserved with clearly defined ownership)
4. Why: • farmers to produce supplies of fuelwood for commercial and subsistence purposes • to supply communities with fuelwood and fodder through community plantations	**4. Why**: • to extend authority to communities to control forest access and allow local management • to regenerate 30–35 million ha of degraded under-productive forest land with regeneration potential • to manage for biodiversity, ecological sustainability and environmental benefits
5. How: • setting budgets and targets • establishing nurseries and plantations • providing employment	**5. How:** • diagnosing social and ecological opportunities • defining rights and responsibilities with respect to products, benefits, protection duties • micro-planning process (negotiation of access controls, silvicultural operations to enhance natural regeneration) • legitimising authority of community management group
6. When: • based on donor aid and budget process • renewal based on target achievements	**6. When:** • based on process of community
7. Average cost: • Rs 5,000–10,000/ha for plantations	**7. Average cost:** • Rs 300/ha for natural regeneration of sal

Source: Arora and Khare, 1994

here that some of the familiar problems encountered with social forestry begin to re-emerge.

In summary, this review of the emergence of forestry over the last 100 years shows a certain recycling of ideas and arguments. It leaves us with many questions as to whether the new approaches herald a new era or a reworking of existing practices (Box 2.11).

Box 2.11 Chronology of Official Responses, 1952–94

1952	Forest policy statement of objectives: increase supply of industrial timber and maximise revenue earnings
1961–6	Third Five-Year Plan acknowledges the need to meet rural energy requirements
1974–9	Fifth Five-Year Plan recognises 'forest and food, forest and people, and forests and wood' as key links
1976	National Commission on Agriculture (NCA) looks into the need to revise forest policy, continues to focus on checking denudation and meeting industrial needs and holds people's privileges responsible for forest destruction
1976	42nd amendment to the Constitution makes forestry a Concurrent subject
1980	Central government abrogates power to decide about diversion of forest land for non-forest uses. An environmental coalition gains in strength, demanding people as the central focus of forest management
1982	Committee on Forests and Tribals recognises the symbiotic relationship between forests and people
1988	Participatory management sanctioned in Forest Policy
1990	Guidelines issued by the Ministry of Environment and Forests on Joint Forest Management
1991	Wildlife Act amended to make the protection of national parks and sanctuaries more stringent and remove all rights by resettling all forest dwellers outside the park boundaries
1994	Draft Forest Act, conflicts emerge between conservation of biodiversity and meeting local people's needs

Source: *Down to Earth*, 1995

Notes

1. This section draws on a paper written by Dargavel *et al.* 1985.
2. These categories are: Reserved Forests; Demarcated Protected Forest Class I; Demarcated Protected Forest Class II; Undemarcated Protected Forest Class III.
3. **Nautor** is an ancient right whereby landless people are allowed to break fresh agricultural land in common land areas. The land is allocated to the landless by village elders, usually on undemarcated (Class III) land.

3 From Forests to Forestry – the Three Ages of Forestry in Nepal: Privatisation, Nationalisation, and Populism

M. Hobley and Y.B. Malla[1]

3.1 Introduction

The history of forest management, degradation and tenure structures throws long shadows over policy and practice today, and indeed still colours local people's perceptions of the intentions of Forest Departments and other outsiders. As will be seen in this discussion of forest history for Nepal, the early period is characterised by growing state control and alienation of local rights, with a concomitant growth in distrust of the motives of the state. Recent changes have led to a reassertion of local rights of access to forest resources, with the government handing over control of hill forests to local users. These changes are discussed further in Chapter 7 in the context of global change and national policies of decentralisation and deconcentration of services from the government sector to non-government and private commercial sectors. Chapter 2 looked at the evolution of the forest sector in India and this chapter traces the emergence of participatory forestry under a different social and political regime in Nepal.

3.2 Privatisation

Forests under the Gorkha Empire (1768–1846)

Before the mid-eighteenth century, present-day Nepal was fragmented into no fewer than eighty small kingdoms and principalities involved in a constant flux of changing alliances (Stiller, 1973). By 1769 the greater part of present-day Nepal was unified into one nation by the King of Gorkha (in western Nepal) Prithvi Narayan

Shah. The period up to 1846 was marked by factionalism within the ruling elite and ended in the dethroning of the Shah kings and their replacement by the Ranas, who ruled Nepal until 1951.

As the Gorkha empire expanded there were inevitable conflicts between it and the British East India Company. The forests of the Tarai, the plains bordering India, were maintained as a physical barrier against possible invasion from the south by the Company (Box 3.1).

The forest resources of Nepal were noted by the British as being of use to meet the expanding demands of its empire. Many accounts report the central role that the Himalayan forests played in providing raw materials for infrastructural development within India (Hodgson, 1972; Bajracharya, 1983a; Tucker, 1983; Dargavel *et al.*, 1985; English, 1985; Guha, 1983).

Forests in the Middle Hill areas were used by local people to support subsistence needs rather than being exploited for external trade. However, as the state and bureaucracy began to grow the revenues obtained from exploitation of natural resources – forests and mining of metals – were formalised and regularised (Mahat *et al.*, 1986). The state asserted its ownership of natural resources and transferred ownership to institutions and individuals as a privilege. These rights were restricted to agricultural and forest lands, and mining rights continued to be held by the state (Regmi, 1984).

The tenure structures that formalised these rights over forests meant that the state did not receive revenues from forests under **birta**[2] tenure and had no direct control over the way in which forests that had been assigned to private individuals were used. However, state-owned forests were protected for the use of the state and in particular to supply the forest-product needs of the ruling families.

The extraction of rents and taxation from agricultural producers forced them to secure their subsistence through cultivation of temporary plots within forests which did not attract any taxes or rents. These plots of **khoriya** cultivation provided millet for the cultivator's consumption. Although there are no records of the amount of land used under **khoriya** cultivation, oral histories from villagers indicate that until recently such practices were widespread and suggest that this practice led to the eventual degradation of large areas of forest.

Although rights over timber extraction were assigned to individuals and certain products from the forests were the right of the

Box 3.1 The Jungle Barrier

In Nepal the duns (the valleys of the inner Tarai region) have been mostly allowed to fall into a state of jungle and are consequently clothed with forests of sal and cotton trees, and are inhabited by wild beasts. The Nepalese are averse to the 'clearing' of these forests, as they look upon the malarious jungle at the foot of the hills as the safest and surest barrier against the advance of any army of invasion from the plains of Hindustan.

Source: Regmi, 1984

state, local people had free access to those forests of limited commercial value for firewood, fodder and medicinal herbs. Hamilton, in 1819, notes that 'in Nepal the pasture and forests are in general common, and any person that pleases may use them' (Hamilton, 1971). However, in some areas of Nepal where forests had a commercial value taxes were charged villagers for meeting their requirements of fodder and other subsidiary forest products (Regmi, 1984). Many tenant farmers working under **rakam** obligations (compulsory labour for landlord) were forced to supply forest products to landowners free of charge.

Privatisation formalised: forests under the Ranas (1846–1951)

Exploitation of certain commercially valuable forests was formalised through the legal-juridical process under the rule of Jung Bahadur Rana (1846–77). A number of rules were passed to regulate access to forests and removal of forest products (Mahat *et al.*, 1986). The passage of these rules coincided with an increased removal of forest products for sale to British India. A forest office was established in Kathmandu to oversee forest exploitation, followed by a forest inspection office (**banjanch goswara**) with a number of check posts to regulate the sale of forest products and the hunting of game (Bajracharya, 1983a).

In the first half of the twentieth century, forest exploitation, particularly in the Tarai region, appears to have increased greatly. At the turn of the century, the British in India extended their railway network to Nepal's southern border, and the Rana Government suddenly found that it could earn more revenue in the region bordering India by clearing forests and producing grain for export (Lohani, 1973). Mahat (1985) has documented how the Ranas established a timber administration office, converted later into the timber export office, and employed British forestry experts from the Indian Forest Service in the 1920s to supervise felling and export of the Tarai **sal** (*Shorea robusta*) forests for the construction of the Indian railways. British influence continued within Nepal through the appointment of a British forest adviser, J.V. Collier (from 1925 to 1930):

How important to the shaping of current Nepalese forest practice and policy was the influence of the British experience in India?

> Government has recently enlisted for a short term of years the services of a British forest officer, who with some fifteen years of experience of the working of forests in India, may be able to induce the best class of Indian contractor to work in the far richer forests of Nepal. (Collier, 1976)

Timber for railway sleepers was granted by the government to the British in India free of royalty as part of Nepal's contribution to the First World War effort (Collier, 1976). The system of forest exploitation remained centred around the use of Indian contractors, and thus it appears that the Nepalese had little control over the exploitation. Indeed, the profits seem to have contributed to the wealth of only three Rana families who between them owned nearly a quarter of a million hectares of Tarai land (English, 1985).

In the hills of Nepal, different systems of forest management operated in conjunction with the **birta** systems: the **talukdari** system and in eastern Nepal the **kipat** system (Fisher, 1989; Bartlett and Malla, 1992). In 1907 an official document (**lalmohar**) providing guidelines for the local use and management of forests through the **talukdars** (local revenue functionaries) was issued. Special decrees or **sanads** were issued to particular landlords whose responsibility it became to manage the forests and organise their protection. Forest watchers or **chitadars** were appointed and paid in kind by the villagers. These systems are reported across Nepal (see Mahat, 1985; Fisher, 1989; Adhikari, 1990; Bartlett and Malla, 1992; Chhetri and Pandey, 1992; Tumbahampe, 1994) and appear to have been practised to a varying extent according to the interests of the **talukdars**. What is interesting from a reading of the translation of the **lalmohar** is evidence of in-built systems of checking and accountability: 'If the talukdar needs timber, he should ask the people and if the people need timber, then they should ask the talukdar' (Adhikari, 1990). The degree to which such accountability was developed can only be the subject of speculation. The **kipat** systems of communal tenure were practised extensively in eastern Nepal and amongst many of the indigenous groups of central Nepal. Under **kipat** local people were allowed to collect forest products, and village headmen (**jimmawals**) were recognised as the tax collectors and became the *de facto* owners of forest lands (Loughhead *et al.*, 1994). The area of forests included under these tenure arrangements declined as they were brought under state ownership by the Rana administration (English, 1982). However, the role of the **jimmawals** as tax collectors and **thulo manche** (big men) continued until the early 1970s. Today they still have a large amount of influence over local matters.

In addition to these systems, **guthi** (religious) forests were also prevalent in parts of Nepal. Here forests were given to a religious institution and the revenue gained from the forest used to support their funds (Tumbahampe, 1994).

It was not until 1942 that a forest service was created within Nepal after another British adviser, E.A. Symthies, who had spent several years with the Indian Forest Service, was asked to advise on the structure of the new forest department. It was formed on the lines of the Indian service, and its forest officers were trained at the (Indian) Imperial Forestry School at Dehra Dun, according to the procedures established for the regulation of Indian forests. Forest exploitation was conducted under a series of working plans, following formats originally established in British India.

At the end of the Rana rule in 1951, at least one-third of the forests of Nepal were under **birta** tenure and three-quarters of this land belonged to the Rana family (Regmi, 1978).

3.3 Nationalisation: Forests under the Shah Monarchy (1951–1987)

Inspired and supported by India, the Nepali Congress revolution of 1951 overthrew the Rana regime and led to a 10 year period of experimentation with democracy, with the reinstatement of the Shah kings as constitutional monarchs. This was a period of great uncertainty which eventually ended in 1960 with the first general election and the release of a constitution for parliamentary democracy. Despite the political and social problems of this decade, some notable legislation was passed which had an impact on the forest sector. The **Birta** Abolition Act was passed in 1959 and much of the forest land previously under this tenure now came within the purview of the state and became subject to tax as agricultural land or was nationalised as forest land.

In 1952–3, a draft forest policy was compiled by Emerald J.B. Rana with E. Robbe (an FAO expert). Bartlett and Malla (1992) show how the rudiments of community forestry were laid down in this policy, but unfortunately were not enacted (see also Gilmour and Fisher, 1991). The following classification of Nepal's forests was proposed (which incidentally looks similar to current classifications and is identical to India's classifications):

1. Protection Forest	Forest which must be preserved or created for physical or climatic considerations
2. National Forest	
i) Forest for Revenue	Forest which has not only to be maintained for the needs of the people, communication and industry, but mainly as a source of revenue
ii) Specific Forest	Forest which has to be reserved for specific purposes like defence, local industries, health and other local needs
3. Community Forest	Forest which has to be created or set aside to provide firewood, small timbers for agricultural implements, building timbers, other forest produce and grazing for cattle, for the rural community

As the extract from the draft policy in Box 3.2 indicates, many of the most important tenets of current community forestry policy are recognised, including the need for partnerships between local people and Forest Departments in order to protect the resource for future generations.

However, this remarkably far-sighted draft policy never entered the formal arena but was replaced with the Private Forests Nationalisation Act 1957 which regulated access to and use of the forests in an attempt to regularise the revenue flow and control of forests in Nepal. The Act needs to be considered in its political

Box 3.2 Community Forestry: Its Origins

As long ago as 1952, community forests were named and described in policy documents. The following extracts indicate the degree to which local people were to be involved in the protection and ownership of local resources. They recognise the partnership role between local people and the Forest Department:

> These forests are intended in the main, to serve the needs of the surrounding villages in respect of timbers for housing and agricultural implements, leaves for manure and fodder, fencing thorns, grazing and edible forest products. These forests shall be conserved or created and conserved by the community around them, under the supervision of the Forest Department. The community are responsible for protection of these forests. Scientific management of such forests for a sustained supply of the villagers' needs will be effected by the Forest Department in full co-operation with the community. The supply for the villagers' requirements should be made available at nominal or non-competitive rates, provided they are utilised by the villagers themselves and not traded.
>
> The management of such community forests should aim at meeting the present as well as the future needs of the population. Removal of the produce in excess of its annual growth should not therefore be permitted. Restrictions should be imposed in the interests not only of the existing generation but also of posterity. The protection of the forests and distribution of produce should be entrusted to *panchayats*. While the profit motive in the management of these forests should be relegated to the background, the expenses for development and maintenance of such forests must come from their own income. Income realised wherever sufficient should be utilised for the community itself.

Source: Bartlett and Malla, 1992

context. The Rana era had ended and the 'democratic' era of the Nepali National Congress had begun. The change in political system led to a call for the end of **birta/jagir** privileges and the repeal of the power of the dominant landowning classes of the Rana period. Incidentally, but perhaps more importantly, the nationalisation of private forests led to an increase in revenue to the state:

> . . . forests constitute an important part of the national wealth and to protect national wealth . . . management and proper utilisation thereof for the public interest, it is expedient to nationalise private forests. (Private Forests Nationalisation Act, 1957)

It defined private forests to bring the tax exempt resource under national control and with it any revenue to be gained from exploitation, though trees in small groups on cultivated lands were excluded:

> Private forests mean all the forests in all land types including

wasteland with wholly or part remission of revenue over which any person is exercising proprietary right (ibid.)

Although the Act appeared draconian, many owners managed to evade nationalisation of their forests and continued to use them for personal gain. Mahat *et al.* (1987) provide a detailed discussion of one such area of forest in Sindhu Palchok where the landowners extracted tribute from the local people in exchange for use of the forest. However, in other cases, interviews with individuals whose families owned large tracts of forest in different parts of the country indicate that in some areas trees were felled to prevent the government gaining control of the land (Hobley, 1990). Local control over forests remained in places where strong local leadership had excluded government interference. In these areas forests were protected through local action to ensure that local people could continue to meet their forest-product needs from the forests, and the Act appears to have had little effect (see Box 3.3 and Loughhead *et al.*, 1994). Indeed, in many cases deforestation appears to have been associated with the cadastral survey which led to the formal demarcation of private land from government forest. Across Nepal studies indicate that these surveys were more influential in hastening the degradation of forests than any piece of policy or legislation (Chhetri and Pandey, 1992; Dahal, 1994; Kafle, n.d.; Karki *et al.*, 1994; Loughhead *et al.*, 1994). The uncertainties over ownership created during these periods of survey led to the opportunist seizure of forest lands and their subsequent conversion to agricultural lands (see Box 3.4).

Although there are many negative experiences associated with survey and demarcation of land boundaries, there have been some positive experiences too. The use of cadastral survey maps has allowed local people to identify areas of community land with potential for bringing under the community forestry programme (VSO volunteer, pers.comm.). Land surveys have brought security of land ownership to many people and allowed user groups to delineate forest land from agricultural land. This has affirmed their

Box 3.3 Local Management Practices

Everyone in the village shared the right to use the forest as needed, but no-one was allowed to clear the land. The village, through its elders and elected headmen, attempted to regulate the amount of cutting . . . To promote a sustained yield, the headmen of the village assign certain rights to gather firewood in certain areas of each woodlot, and households jealously guard their territories; many territories represent traditional claims that date back several generations . . . (W)hen a household needs a particularly large tree for a construction project, they must pay a sizeable sum to the village headmen . . . (T)he headmen will sometimes declare a moratorium on cutting if a certain plot shows signs of really excessive use that will soon lead to complete exhaustion.

Source: Cronin, 1979

Box 3.4 From Common to Privatised to Open-access Property

Hark is over 50 years old. According to him, the Ramche Ban originally belonged to the clan of Mr Tirth Raj Bhattarai who claimed that the forest was gifted to his ancestors by the then **jimal/talukdar** during the Rana regime. After Tirth Raj's death, his son, Dibakar Bhattarai, controlled the forest. The Bhattarais were very strict in enforcing the communal forest rules. Without the permission of the Bhattarai brothers, nobody could enter the forest. Felling of big trees was allowed only in cases of dire need such as fire damage to homes and other destruction wrought by natural calamities. When news of the nationalisation of forests, abolition of the **talukdari** system, and introduction of a cadastral mapping system reached the village, the Bhattarai family lost control of the forests. All the people of Rampur and the adjoining villages started encroaching upon the forest. By 1969, the original forest cover was completely destroyed. Some settlers also moved inside the forest area. The whole landscape looked deserted and only undesirable shrubs were left to grow. The naturally regenerating vegetation was also harvested quickly. By 1982, one could see birds walking on the terrain.

Source: Kharki *et al.*, 1994

In 1957 the forest was nationalised by the King but the people living in the remote hill areas did not know about this. I also did not know until much later. We continued to preserve the forest in the same way as we had since 1951. (Laxman Dong, 1986)

(Hobley, 1990)

right to assert control over the maintenance of these boundaries and to prevent encroachments on forest margins.

In 1959, the first Forest Ministry was established to serve the entire country. However, at this stage, there were still very few trained staff and thus management of each patch of forest was not possible. The hill forests were not brought under any form of working plan. Overall, forests remained unmanaged in the formal sense and the forest administration understaffed and underdeveloped.

In 1960 the democratic experiment came to an end with the expulsion of the elected government by King Mahendra on charges of corruption (Joshi and Rose, 1966; Gupta, 1994). In 1962 the King introduced the **panchayat** system and political parties were banned. Thereafter followed 30 years of relative stability in which opposition was firmly controlled, until 1989 when the first stirrings of dissent broke out.

3.4 Development of Forestry Institutions

The development of the Forestry Department in Nepal occurred relatively late compared with India, and is a reflection of the form of rule adopted by the Rana regime. Forests in the Hills were of minimal revenue value to the regime and were of better service if converted to agricultural land (compare this with the similar arguments made in India). The valuable forests of the Tarai were in the main inaccessible to exploitation and generally the climate was inhospitable to outsiders, with malaria a widespread and deadly problem.

As can be seen in Table 3.1 it was not until the late 1960s/early 1970s that the Forest Department began to take shape as a recogni-

Table 3.1 Development of the Forest Department

Date	Administrative change	Function
before 1927	No administrative forest offices	Distribution of lands under **birta** tenure
1927	**Kath Mahal** (forest office) established	To supply railway sleepers to India
1939	'Eastern Wing' and 'Western Wing', established	To supply sleepers and collect revenue
1942	DOF established with 3 circles and 12 **Ban Janch**	To control and manage forest administration
1951	2 circles and 44 ranges covering the Tarai areas. IOF established	To control and manage forest activities in Tarai areas
1959	MOF established	To cover forest activities nationwide
1960	MOF abandoned (lack of staff). CCF Office established with 7 circles and 22 divisions, covering some hill areas	To collect revenue. External assistance started
1961	Timber Corporation of Nepal established	To utilise timber from resettlement areas in Tarai
1962	Working plans prepared for some Tarai Districts	To start planning process in forestry activities
1966	Fuelwood Corporation established	To supply fuelwood to Kathmandu
1968	14 circles and 75 DFOs (but failed due to lack of trained manpower)	To coincide with development of other administrative structures
1968	7 circles 22 divisions, and Pradhan Ban Karyalaya	To strengthen the organisation with available manpower
1976	National Forestry Plan published 9 circles, 40 divisions, covering 75 districts	To implement forestry activities nation-wide on a planned basis
1983–1988	5 Regional Directorates and 75 DFO offices	To match Decentralisation Act

Source: Joshi, 1993

sable public sector entity with clear territorial jurisdiction.

Following the failure of the democracy movement and the reassertion of monarchical rule in the early 1960s, a new partyless panchayat system was introduced, which was to remain in place until 1990. Together with the Forest Act of 1961, this had far-reaching consequences for the local control of resources. The Act included a provision for handing over the protection of forests to the newly formed panchayats (a provision which was adopted by several panchayats, particularly those close to Kathmandu). Several categories of forest were delineated, each with different rights of access assigned to it. Here we see a further breakdown of the category of 'community forest' now renamed as panchayat forests:

- Panchayat forests: any government forest or any part of it, which has been kept barren or contains only stumps, may be handed over by HMGN to the village panchayat for plantation for the welfare of the village community on the prescribed terms and conditions

- Panchayat Protected Forests: government forest of any area or any part of it may be handed over to the panchayat for protection and management purposes (this generally referred to existing natural forests and not to 'blanks')

- Religious Forests: government forest located in any religious spot or any part of it, may be handed over to any religious institution for protection and management

- Contract Forests: any government forest area, having no trees or only sporadic trees, may be handed over to HMGN in contract to any individual or institution for the production of forest products and their consumption.

Ownership of the forest land remained with the government and control could be resumed whenever the government deemed it necessary. The panchayat had some powers to fine those who transgressed the law. However, management decisions remained with the government forest service. Private forests that were considered to be poorly managed could be taken over by the government for a period of 30 years, and any income from the forest would be given to the owner less a sum deducted for management costs. The Forest Act of 1961 legitimised panchayat managed forests but not forests managed by the users. This Act, although it laid some of the foundation for future community forestry policy, had little impact on those areas distant from Kathmandu where local people continued to use the forests for their subsistence needs, under locally derived regulatory frameworks, regardless of the national legislation.

Although between the early 1960s and the mid 1970s there appeared to be little outside interference in the forest sector, the development of the forestry service remained under the control of an expatriate Chief Conservator of Forests, R.G.M. Willan, until 1967. Internal control at a national level was not assumed until the late 1960s.

Like India, Nepal introduced more punitive forms of forest policy. In 1967 the Forest Preservation (Special Arrangement) Act was passed. This defined forest offences and prescribed penalties for them including the provision to 'empower district forest officers and guards to shoot wrongdoers below the kneecap if they in any way imperilled the life or health of forest officials' (Talbott and Khadka, 1994), thus confirming the role of the nascent forest service as one of policing and licensing. This Act together with several other land-related Acts (Birta Abolition Act (1959), Land Reform Act (1964) and the Pasture Land Nationalisation Act (1974)) increased the power of the Forest Department and extended government control to all areas of land outside private cultivation and ownership. However, looking back at Table 3.1 it is apparent that, despite all these powers, the size of the Forest Department rendered it ineffective in most remote hill areas (Gilmour and Fisher, 1991).

The global justification for forestry for industrialisation was vigorously pursued in Nepal with a period of heavy exploitation of the Tarai forests. In the 1960s the government encouraged the establishment of both large and medium-scale forest-based industries, and established the Timber Corporation of Nepal (TCN) to supply timber to these industries as well as to urban centres for construction (HMGN, 1990; Sheikh, 1989). At about the same time, the government also established the Nepal Fuelwood Corporation to supply fuelwood to urban centres, with supplies largely coming from the Tarai (Sheikh, 1989). Since the major source of revenue came from the extensive and easily accessible Tarai forests, all the forestry infrastructure was concentrated there with few personnel posted to the hill areas. Thus the hill forests remained relatively untouched by bureaucratic procedures and were not brought under any form of systematic management. This fundamental difference between the exploitation history of the Tarai and hill forests continues to influence the development of participatory forestry in these two regions.

3.5 The Emergence of a Populist Forestry

The next push towards community forestry came from an indigenously identified need in 1974. During this year the Ninth Forestry Conference was held in Kathmandu, which convened forest officers from all over Nepal. A community-oriented group of foresters working in the districts met to promote a new form of forestry based on their experiences of working with local people in forest management. They called this type of forestry community forestry, and attached practical experience to what had been policy rhetoric until this time (T.B.S. Mahat, pers. comm quoted in Hobley, 1990).

The proceedings of this conference formed the basis of the 1976 National Forestry Plan which re-emphasised the rulings of the 1961 Forest Act in allocating categories of forest land to the panchayats. It also recognised the shortcomings of forest policy and stated that 'the Forest Department had been ignoring the forests in the hills region and this has led to the deterioration of the watersheds which are now in very poor condition' (NAFP, 1979). Under the Plan, wider powers were given to District Forest Officers to formalise the transfer of nationalised forest land to panchayat control. In 1978 the Panchayat Rules were promulgated, which then provided a framework for the operation of a new fleet of externally funded community forestry projects. International donors poured funds into Nepal on the basis of 'saving the environment' from further degradation.

Even at this relatively late stage, the formal government institution responsible for forests was still underdeveloped. For example, in 1976 forests in remote areas were still not under the control of the Forest Department but remained the responsibility of the highest government officer in each district. The Forest Department was reorganised after 1976 to ensure that each district's forests came under the management of a forest officer. The staffing at field level remained relatively low, which meant that management of forests could only be implemented through a strictly enforced protection policy using fear to regulate forest use. However, as has been said, in most areas there was no government presence and forests were controlled by the local people under a variety of indigenous management systems. Throughout this time therefore the Forest Department's role was entirely custodial with no active management of forests. Perhaps this reveals the inadequacy of the rigidly applied Dehra Dun model (as discussed in Chapter 2) originally established by the British, and its attempted implementation in a situation where the necessary physical and bureaucratic infrastructure was absent. Throughout this time local people were forced to use the forests illegally to meet their basic firewood and fodder needs, and in many cases, as is discussed below, access to these government lands was regulated by local practice and not by Forest Department staff.

The eco-doom scenario 1970s–80s

As international attention was drawn to the floods in Bangladesh and their link with deforestation in Nepal, the prescribed solutions became relatively simple – plant more trees. Thus by the end of the 1970s, the international community considered Nepal to be facing an ecological, social and institutional crisis of enormous proportions which would have far-reaching consequences for other countries in the region, in particular Bangladesh (Hoffpauir, 1978). With interna-

Community forestry nursery, Nepal

tional attention focused upon it, Nepal entered the 'eco-doom era'. To understand the context in which community forestry claimed such international attention, it is important to return to one of the most influential eco-doom writers of this time – Eric Eckholm. Ives (1987), in a useful detailed account of this period, provides a summary of the eight major points of what he called 'The Theory of Environmental Degradation' (see also Ives and Messerli (1989)). These are presented in Box 3.5.

These powerful arguments served to fuel an already alarmed international debate and informed the 1978 World Bank review of the Nepal forestry sector in which figures and predictions showed how the landscape of Nepal, through the acts of hapless peasants, would be reduced to a barren wasteland. Griffin (1988) provides a concise summary of the main conclusions of this alarmist report which was to form the basis for all major funding programmes for the next decade (Box 3.6) and indeed continues to cast long shadows over much current practice.

What myths still remain unchallenged in current forest policy and practice in Nepal? Consider the role of livestock in forest-farming systems; analyse the myths surrounding their role in the degradation of forest resources.

Many of the assumptions underlying these statements still remain unchallenged and continue to inform policy and practice, and indeed it is difficult to provide the empirical evidence to support or contradict them. Although work in Nepal over the last 15 years has allowed many of these myths to be challenged (see Donovan, 1981; Bajracharya, 1983a, b; Arnold and Campbell, 1986; Nield, 1985; Mahat *et al.*,1986 and 1987; Hamilton, 1985; Thompson *et al.*, 1988; Griffin, 1988; Fox, 1993) many reports, particularly in the media, still cite the peasant as the main cause of degradation, neglecting the institutional and social framework within which that individual operates and which may be equally responsible for the degradation (Box 3.7).

The confirmation that there is, in fact, land degradation taking place can be a daunting task. The momentum of government publications, received wisdom and academic research can so condition the perceptions of policy-makers at a point in time that it is sometimes difficult for any counter-intuitive results of research to gain credibility. The underlying reason for this is that, in spite of all advances in data collection, remote sensing and basic research into the physical processes of land degradation, there is usually insufficient evidence that irreparable damage is taking place.

(Blaikie, 1990 quoted in Hausler, 1993)

Thus it was into this climate of doom that community forestry was introduced and the international donor community began the process of shaping a forest sector that had been underdeveloped since the 1960s. It was also a time when the Forest Department recognised that it had 'neither been able to stop the destruction of

Box 3.5 The Eight Steps to Disaster

1. Following the introduction of modern health care and the suppression of malaria after 1950, the Nepalese population began to grow rapidly.
2. The consumption of forest products correspondingly increased and much forest was converted to agricultural land.
3. Excessive pressure resulted in massive deforestation.
4. Deforestation and the expansion of agriculture on to ever steeper slopes resulted in catastrophic erosion.
5. Increased run-off and siltation cause severe flooding at lower levels.
6. The increased sediment load of the rivers results in significant extension of the river deltas, and the creation of new islands in the Bay of Bengal.
7. Shortage of fuelwood leads to the increased burning of dung.
8. Shortage of natural fertiliser causes decline in agricultural productivity and the attempt to convert yet more forest to arable production.

Source: Ives, 1987 quoted in Griffin, 1988

Box 3.6 Whither the Trees and Soil?

Increasing population was seen by the World Bank as causing both a direct reduction in forest area because of conversion to agriculture and a rapid degradation of the remaining forest. The combined effect was stated to have been the complete destruction of about 1,000,000 ha of hill forests in the decade before 1978. All the accessible forests in the hills would disappear by 1993 and in the Tarai by 2003 unless large scale compensating action was undertaken. About 1,300,000 ha of plantations would be needed by the end of the twentieth century. In the absence of such compensating action, rural people would be forced to burn ever-increasing amounts of dung and agricultural residues. The consequent decline in the fertility of arable land might lead to foregone grain production of over 1,000,000 tonnes per annum before the end of this century. The role of the forestry sector in underpinning the quasi-subsistence agricultural economy was emphasised. Forest products were used as fuel, fodder, poles and other building materials . . . The Review outlined the links between forests, fodder, livestock and the productivity of cultivated lands and stressed that the linkage was fundamental to any properly based strategy for forestry in the hills. Livestock grazed and browsed both forest and agricultural land . . . A reduction in livestock numbers would be beneficial but was unlikely to occur because of social and religious attitudes . . . The effect of erosion was also considered to be of great significance. Half the erosion in the hills was associated with human activity. High population density, the cultivation of steep slopes without appropriate conservation measures and excessive livestock numbers were all important contributors to erosion. . .

Source: adapted from Griffin (1988)

the forests nor been able to manage the remaining forests in successive years' (Joshi, 1993).

The decade of the 1980s was an extraordinary period of experimentation with different forms of community forestry and unprecedented donor activity, with each donor adopting a different district and implementing its own interpretation of what constituted best community forestry practice. Although at times different practices caused conflicts, on the whole this cauldron of ideas allowed the emergence of a form of community forestry highly suited to the particular needs of the hills environment. The Government of

Box 3.7 Disaster Hits the Environment

The claim that the nationalisation legislation led villagers to feel that the government had taken their forests away is an oversimplified view. It would have been almost impossible, more than 30 years ago, to have conveyed the intention of the government, expressed through legislation, to the villagers in rural Nepal. A more powerful influence has been the recent activities of land survey teams. Once villagers know that the government is going to fix the boundaries of land holdings, they often scramble to claim as much area as possible for themselves prior to the arrival of the survey teams. This often results in the clearing of trees and shrubs and the start of cultivation on land that may have been previously used to harvest forest products.

Source: Kayastha, 1990

Nepal's forest sector policy was first declared in the Sixth Five-Year Plan (1981–5) which emphasised community participation in the management, conservation and utilisation of forest resources.

This array of diverse experiences gained its greatest focus in 1987, when the government undertook the task of developing a 20-year Master Plan for the Forestry Sector (MPFS). This placed greater emphasis on community forestry, with 47% of proposed investment to the forest sector in support of community forestry programmes.

The Master Plan formed the basis for a draft forest policy in 1989, the first priority of which was to meet the basic forest-product needs of local people through community forestry and private planting. Several principles were clearly articulated to meet this priority:

- phased handing over of all accessible hill forests to the local communities, to the extent that they are willing and able to manage them
- entrusting the users with the task of protecting and managing the forests and receiving all the income
- emphasis on an extension approach aimed at gaining the confidence of the woodcutters and others, particularly women, who actually make the daily management decisions
- retraining the entire staff of the Ministry of Forests and Soil Conservation for their new role as advisers and extensionists.

These changes represent a significant shift away from the objectives of the 1950–70s of generating revenue for the state by exploiting forest resources. Unlike the past policies which concentrated forestry activities in the Tarai and the urban areas, the new policy document puts the emphasis on a more geographically holistic approach. This requires major changes in the institutional and legislative framework, which are being implemented in the wake of the recommendations of the MPFS.

In order to understand the impetus for this major change in policy, it is necessary to consider a series of events both internal and external to Nepal. Although one of the most important internal events was the Ninth Forestry Conference, this was perhaps the catalyst for a series of other equally far-reaching changes, including the passing of the Decentralisation Act in 1982. The Act formalised the duties and responsibilities of village panchayats and ward committees, and empowered them to form:

> people's consumer committees to use any specific forest areas for the purpose of forest conservation and, through them, conduct such tasks as afforestation, and forest conservation and management on a sustained basis. (HMGN translated by Regmi, 1982)

The Decentralisation Act and Rules passed in 1984 went beyond the original Panchayat Forest Rules which designated the village panchayat as the local institution for forest management. A 1988 amendment to the Panchayat Forest and Protected Forest Rules of 1978 subsequently adopted the concept of the user group by making reference to the Decentralisation Act.

The Seventh Plan too (1985–90) was explicit in its support for people's participation. One of its main policies was stated as:

> To mobilise people's participation in massive afforestation and forest conservation programmes in order to ensure supplies of essential forest products to the people. (National Planning Commission, 1985)

Community forestry implies . . . any situation which intimately involves local people in a forestry activity exceeding the mere payment of wages for labour, regardless of ownership of the land . . . It embraces a spectrum of situations ranging from the production of forest products for local needs, the growing of trees at the farm level as a fodder crop, the processing of forest products by the household, artisan or small industry to generate income directly or indirectly, to the activities of forest-dwelling communities. It includes activities of forest industry enterprises and public forest services which encourage and assist forestry activities at the community level.

(NAFP, 1985)

1987 was an important year for the development of community forestry, when policy makers, Forest Department staff and project staff came together for the first National Community Forestry Workshop. Recommendations from this workshop included acceptance of the user group concept which was later incorporated in the Master Plan. Thus the stage was set: an enabling framework was in place, experience was being developed and community forestry was being funded by government and donors alike. Supported by increasing awareness of local people's knowledge and experience of growing, managing and utilising their forests and tree resources, this has been fundamental in the development of a more participatory form of forestry in the 1990s in Nepal.

The government-supported community forestry programmes in the initial years of the 1980s were based on the assumption that the major cause of deforestation in the country was illicit cutting and clearing of forests by 'short-sighted', 'uneducated' and 'ignorant' villagers (Hausler, 1993). One of the solutions was seen to be 'teaching' villagers about the importance of forests and trees and 'motivating' them to plant and protect forests. This led to a large programme of reforestation, with browse-resistant species, of panchayat and government lands, usually those identified as 'barren',

Roofing shingles cut from high altitude forest, Nepal

again mirroring the debate about 'waste' in India. In the main, decisions were taken by a few of the local leaders on behalf of the local people, often with no consultation with those who were actually using these lands. Many of the problems experienced with this form of community forestry are those already discussed with regard to social forestry. Indeed, parallels can be drawn between community forestry and user group forestry similar to those between social forestry and joint forest management. The major difference is that in the former types of forestry it is the trees that are paramount and in the latter the people and the managing institution (Gilmour and Fisher, 1991).

As project and government staff gained more experience there was a more general questioning of the underlying causes of deforestation. Several projects reappraised their interpretation of community forestry and began to look in detail at the communities and their existing forest practices. The evidence provided by several studies suggested that farmers are not ignorant but are quite capable of managing their natural resources. Farmers have not been wantonly destroying forests and trees, but in many cases have preserved and planted trees on their private lands without any outside support (Campbell, 1978; Molnar, 1981; Messerschmidt, 1981; 1984; 1986; 1987; Pandey, 1982; Acharya, 1984, 1989; Campbell *et al.*, 1987; Robinson and Neupane, 1988; Byron and Ohlsson, 1989; Fisher, 1989, 1991; Fisher *et al.*, 1989; Malla *et al.*, 1989; Carter and Gilmour, 1989; Gilmour *et al.*, 1989; Hobley, 1990; Jackson, 1990; King *et al.*, 1990; Gilmour and Nurse, 1991; Gilmour and Fisher, 1991; Carter, 1992; Chhetri and Pandey, 1992; Gautam, 1992; Karki *et al.*, 1994). This led to a major reorientation of practice in which projects, together with Forest Department staff, began to support local-level management of existing government-owned forests. This was a fundamental shift from panchayat or village-based

Irrigated land is usually managed without trees in the hills of Nepal

forest management systems as the unit for organisation to user group forestry.

The emphasis on user groups rather than panchayats or village development committees emerged from experience gained from the implementation of community forestry between the mid-1980s and 1990. The user group as an organising concept was formalised post-1990 in legislation and policy statements.

Despite the enormous amount of energy and commitment from government staff and donor organisations, by 1987 only about 2% of the available local forests had been handed over to local management, and community forestry still did not seem to be more than a minor addition to Forest Department practice (Talbott and Khadka, 1994). In many cases, local people still considered the plantations created through their labour as government forests and were generally unaware of the nature of community forestry (J.C. Baral, 1993).

The following case study looks at the development of local forest management in one area close to Kathmandu. It highlights the importance of understanding the history of forest management and questions assumptions, which were prevalent in the 1980s and formative of project practice, about local people's abilities to manage resources.

3.6 What Causes Deforestation: A Case Study from Central Nepal

Deforestation has a long history which predates the living memories of the older men and women in the villages making up this Village Development Committee. In their memories the high ground was maintained as grazing land, and was thought to be a remnant from the period of Prithvi Narayan Shah and his capture of the Kathmandu Valley in 1768, when trees were cut down for firewood to enable the armies to obtain a clear view into the Valley.

From 1939 AD (1895 BS), oral histories suggest that there was increased encroachment on forest land. People began to cultivate grazing and forest land adjacent to their fields:

> We call this land 1895 BS because after this date the land was registered ... People would cut down trees next to their fields for firewood. Next year they would include this forest land in their cultivated area, and so they would go on increasing their land. (Hari Chhetri quoted in Hobley, 1990)

During the Rana period rights over the use of forests were vested in the government through the **ban goswara** (forest survey office) in Kathmandu and implemented through a **dittha**:

> The **ban goswara** used to have the overall authority and ... the **ditthas** were employed from the **goswara**. Government forests

> were demarcated by the **ditthas** using stones. Anyone who was powerful could get the post of **dittha**. People went to the **goswara** office and applied for the post of **dittha**. The boss of the **goswara** had the right to employ **ditthas** and he would employ his own people. The villagers used to be the guards and they were paid by the Ranas ... They had the same responsibility as the guards of these days have. (Hira Shrestha quoted in Hobley, 1990)

The **ditthas** and guards had jurisdiction over 7 or 8 forests in the area but were unable to enforce the restrictions because they visited each forest only once a month. However, since the forest resource was extensive at this time and populations relatively low, there was little pressure on the forests. During the Rana period most forests in the village area had restrictions on their use. For example, dry leaves could only be collected once a year during mid-March to mid-April, and access to the forest was prohibited at any other time by the village elders. Firewood that had been cut in previous months by woodcutters, assigned by the elders, and left in piles to dry, was then made available to the villagers in the same period. Villagers from settlements other than the local village were forbidden access to the forest, and if they were found violating this rule they were punished by confiscation of their baskets and cut firewood. Local mechanisms existed through which serious disputes could be resolved. For example, meetings between aggrieved parties decided that cut timber should be confiscated and put in a public place for communal use by the villagers who were designated users of the forest from which the timber had been removed.

Deforestation: 1934–50

Large-scale deforestation of these forests occurred during the Great Bihar Earthquake of 1934. This was followed in 1935 by a heavy snowfall which further exacerbated the damage to the forests. The earthquake caused widespread destruction of houses and forests. The three towns of Kathmandu valley were devastated (Bilham, 1994). As a result of the damage an order was issued by the government opening forests to local people to supply timber for rebuilding. However, the contractors were said to have used this order as an excuse to mine the forests for timber and were responsible for felling large areas of unbroken trees to be sold in Kathmandu to repair damage sustained there.

Freeing the access to forests led to further destruction, with many people cutting and selling firewood to the neighbouring towns (Hobley, 1990). Government controls could not be sustained when demand for forest products outstripped supply, and forests close to trails rapidly became degraded. Oral histories from the villagers indicate that from the 1930s they were forced to travel long distances to other forests for firewood, fodder and timber:

> At that time (1930s) there were forests about 6 miles east of here. Women had to walk 12 miles a day to get one **bhari** (large basket) of firewood. They left home early in the morning and came back in the evening with a bundle of firewood. (Ram Chhetri, quoted in Hobley, 1990)

The hillside now dominated by Sano Ban was then bare of all trees and was used as village grazing land.

> Democracy came to Nepal in 1951. The destruction of the forest also began with the beginning of democracy' (Hari Prasad Bahun quoted in Hobley, 1990).

During the Rana regime if people were found cutting down trees they were fined; after the introduction of democracy the rules were relaxed and people reportedly bribed the forest guards to gain access to the forests. It was a period of increasing population with insufficient food production, and firewood and timber were cut for sale in the towns to provide cash to buy grains. This period also saw the afforestation of grazing land above the village. Thulo Ban was planted by the Afforestation Division with species of pine. Local people indicate that they were opposed to the total loss of the communal grazing area and asked the government to leave some land bare for the cattle to graze on. The impact of national legislation on forest use in this village area was minimal. Villagers did not know about the Private Forests Nationalisation Act of 1957, and did not consider that there was increased deforestation at the time of the introduction of the Act.

The hillside which was formerly grazing land is now covered with trees, Nepal

The Shah monarchy: 1960–89

There was no significant increase in forest cover until the introduction of the panchayat system in 1961–2. Major expansion in cultivation on to grazing lands in the higher regions of the panchayat is a relatively recent occurrence. Land which is now afforested and terraced on the upper slopes of the panchayat was grazing land until 1963.

Carrying pine needle litter for use as animal bedding, Nepal

The advent of the panchayat system and the delegation of responsibility for local resource management to the panchayat led to a significant change in forest usage patterns. Upland forests that had been open to the villagers were now protected by surrounding villages and the villagers therefore had to look to alternative resources to supply their needs of firewood, fodder, leaf litter and green bedding. The decision was taken by the village elders to protect forests that lay within their ward boundaries and to exclude outsiders. Each ward was to have access rights and control over forests within its boundaries:

> With the administrative division of the panchayats, the forests were divided. The forest that lies in a certain panchayat area belongs to that panchayat. The panchayats take care of the forests that are within their boundary. Previously it was not like this – forests were free to all and people could go to any forest. (Bahadur Chhetri, pers. comm.)

The village elders formed a forest committee, appointed a forest guard from the village and made the villagers sign a paper to say that 'they and their descendants would go to hell if they destroyed the forest' (Maya Chhetri, pers. comm.). Through the controls instituted by Chettrigaun, 25 years later Sano Ban is no longer a bare hillside but a dense forest supplying the villages with firewood, fodder and animal bedding:

> Many women and men go to the forest. People used to be allowed to collect dry leaves only between Phagun (mid-February to mid-March) and Baisakh (mid-April to mid-May). The elders used to fix one day during Baisakh when all the villagers could collect dry leaves. On that day all the family would go to the forest and collect leaves, even children and men. The men collected the leaves and the women carried them home. On that day people would also hire labourers to collect and carry leaves. (Sati Chhetri and Sangari Chhetri).

Different settlements adopted their own forest protection practices but they all employed a villager to act as a forest guard. This is described in Box 3.8.

Local forest protection systems continued for between 15 and 22 years in this area, ending in 1982. Conflicts between individuals led to the eventual demise of the systems and ultimately to the total degradation of one of the forests. Cases were cited of forest guards being bribed by wealthy villagers, and in other areas of unregulated stealing from the forest during the night: 'People claimed that the guard did not treat everyone equally . . . he helped those who were powerful' (Ram Chhetri, pers. comm.).

In parallel with local systems of protection, government forest guards were appointed to protect these forests. However, these guards had limited effect; forests were considered to be the prop-

Box 3.8 The Mana-pathi System

People started to take care of the forest after the panchayat system was introduced. They employed a forest guard. The guard was paid from donations collected by the villagers. Not all the villagers have trees in their fields; people who do not have trees of their own steal firewood from the forest, even these days. The guard was paid in food grains and he could sell the grains if he needed money. . . He was paid 2–3 pathis of grain by each household. He was paid 6 manas per person. Children below 16 and people over 40 were not counted. So the quantity that a household had to pay depended on how many family members there were between 17 and 39.

This system continued for 15–20 years. It has not been in practice for the last 3–4 years. It was stopped because of a quarrel. The villagers quarrelled with the guard. People wanted to cut firewood. The guard did not let them so there was a quarrel. The guard gave up his job; he said he was looking after the forest for the welfare of the villagers, not for his own interest.

In neighbouring villages similar systems were in operation. The ruling was so rigidly enforced that for some poorer households it was difficult to maintain payments. This resulted in their borrowing grain from other villagers to pay their dues.

Source: Bahadur Chhetri and Thulo Sunaar, pers. comm.

erty of particular villages and were protected by them irrespective of the presence of a government guard: '. . . for old people like us the government forests, private forests and panchayat forests are the same. We can take care of the forests whosoever's forest it is' (Raj Bahun, pers. comm.). However, it was also the case that forests that had no local protection system did rapidly degrade and were effectively treated as open access resources.

This brief tour through the history of one forest area shows the complexity of the reasons for deforestation and how what at first sight could be considered to be a downgrading forest could in fact be an upgrading forest. As Appendix E indicates, before starting any new forest management intervention in an area, it is important to understand the history of forest management, and in particular to understand the reasons for the apparent deforestation. It may not be the usual simple causal relationship between too many people and too few resources, and therefore interventions should also be tailored to reflect complexity rather than simple solutions that may solve only apparent problems.

3.7 The Language of Participation: the 1990s

1990 heralded the beginning of a new era for Nepal and in consequence a new era for the forest sector as well. Just as India had struggled with the implementation of social forestry which was top-down, prescriptive and target-driven so too Nepal found that the form of community forestry practised during much of the 1980s was not really developing good systems of local management. The changes brought about after the National Workshop in 1987 provided the foundations for new legislation to be enacted in 1993.

In March–April 1990, the banned Nepali Congress and several Communist parties launched a joint movement demanding the restoration of democracy based on political parties. Following a period of violent civil unrest, King Birendra agreed to disband the panchayat system, and allow political parties and to remain as the constitutional monarch. A new Constitution was promulgated in 1990, which made provision for a new tier of local government structures stretching down to the former village panchayats and replacing them as the lowest unit of administration with village development committees. However, there was no real change in their geographical boundaries. In 1991, general elections followed and a democratically elected Congress Government came to power.

The legal framework under which community forestry now operates has three principal guiding documents: the revised 1989 Forest Policy of the Master Plan for the Forest Sector; the Eighth Five-Year Plan; and the Forest Act of 1993 (see Box 3.9). Although two of these documents were constructed post-1990, they share many similarities with earlier legislation and policy. Just as in India, there are contradictions between policies, legislation and practices concerning the decentralisation of control to the local level. As Talbott and Khadka (1994) suggest, the top-down panchayat system and the remnants of the Rana feudal period still penetrate all aspects of the forest sector.

Although the new Act acknowledges the rights of user groups to manage and protect forest areas, it also states that ownership remains with the government, which retains the sovereign right to take back possession of the community forest if the terms and conditions of the handover are not met. The new legislation gives

Box 3.9 The Forest Act of 1993

The Forest Act of 1993 repeals the panchayat forest legislation of 1961 and 1967. The by-laws provide the legal bases for implementation.

The Act acknowledges the same five categories of national forests formally established during the panchayat period:

- community forests that are entrusted to *user groups* for management and sustained utilisation
- leasehold forests on land that has been leased by central or local authorities to individuals or groups
- government-managed forests in which production forest units are managed by a centralised government system
- religious forests belonging to religious institutions; and
- protected forests

The land is still owned by the national government. In the case of community, leasehold, and religious forests, the respective community users' group, lessee or religious institution owns the trees.

Source: Talbott and Khadka, 1994

unlimited powers to the DFO to control user groups managing forests, with little protection for the users in case of a dispute between them and the Forest Department. There are still many contradictions in the Act which are apparently the result of its being drafted during a period of great social change (between 1990 and 1992) (Talbott and Khadka, 1994). However, many of the problems within the legislation are the same as those emerging in India and will be discussed in greater detail in Chapter 5. Nevertheless, as Table 3.2 indicates there has been significant progress in bringing forests in both the hills and the Tarai under user group management. Recent figures (February 1996), drawn from the Community and Private Forest Division database of the Department of Forests, indicate that these figures have now risen to over 3,000 forest user groups managing an area in excess of 200,000 ha.

The apparent ambiguity between lenient practice in the field and actual legal power does lead to some dilution in the rights of local people. However, the Act is still a progressive piece of legislation which permits the following activities (Shrestha, 1995):

- authority for handing over forests to users has been devolved to DFOs
- surplus income generated from user group-managed forests can be used for development other than forestry
- the users are responsible for drawing up operational plans
- users can fix the rate at which forest products are sold, irrespective of government royalty rates
- community forestry retains priority over other national forestry programmes; and
- forestry user groups can register themselves as independent bodies

The recent finalisation of the Forest Rules (1995) through which to implement the Forest Act has further clarified the powers of forest user groups, and has signified a fundamental change in community forestry, where user groups are now allowed to establish their own

Table 3.2 The number of user groups formed and the area under community forest in Nepal

Region	No. of forest user groups (December 1994)	Area under community forest (ha)
Mountain and hills	2,489	93,491
Tarai	267	19,135
Total	2,756	112,626

Source: Shrestha, 1995

Table 3.3 The emergence of participatory forestry in India and Nepal: 1800s to 1990s

Period	Form of government	What happened to the forests
1800–1850s	**India** East India Company and colonial administration	• Conversion of forest lands to agriculture • Extraction of timber for Company needs • Assertion of colonial control over forest lands
	Nepal Shah monarchy	• Gift of forest lands in lieu of payment for services to state mainly in the hills
1850s–1920s	**India** Colonial administration	• Exploitation of forests to meet infrastructural needs • Settlement and exclusion of local user rights • Revenue maximisation • Conservation forestry
	Nepal Rana feudal regime	• Gift of tax-free forest lands in lieu of payment for services to state, practice extended to Tarai areas • Extraction of Tarai timber to supply to British India
1920s–1960	**India** Colonial administration to independent India	• Revenue maximisation • Industrial forestry • Nationalisation • Local protest against forest policies
	Nepal Rana feudal regime to democratic government to Shah monarchy	• Timber exports • Clearance of forests for resettlement • Nationalisation
1960–70	**India** Democratic government	• Industrial plantation forests replaced natural forest • Increasing protest against forest policies
	Nepal Shah monarchy	• Industrialisation • Provision of panchayat forests for local control • Protection-oriented laws enacted • Continued resettlement in Tarai forest areas • Continued timber exports

Table 3.3 (continued)

Period	Form of government	What happened to the forests
1970–1980s	**India** Democratic government	• Increased central control over forest land allocation • More stringent forest conservation rules • Introduction of social forestry on non-government lands ('wastelands') for fuelwood and fodder • Increased voice of environmental NGOs
	Nepal Shah monarchy	• Eco-doom forestry • Introduction of community forestry on non-government and government lands – plantations for fuelwood • Continued timber exports
1980s–1990s	**India** Democratic government	• National forest policy favouring participatory forestry • Joint Forest Management Resolution for management of government forest lands • Voices of biodiversity and industry lobby increasing in volume • Draft Forest Bill reflects internal conflicts between old type of forestry and new participatory devolved management forms • Variable percentage share of benefits between government and forest users • Government retains all resource rights to Reserved and Protected forests • Land tenure remains with government • Government retains right to reclaim forests if misused by local people
	Nepal Democratic government elected with constitutional monarchy	• New Forest Act emphasises importance of participatory forestry, still some fundamental contradictions • Emphasis on user groups rather than administrative definitions for local organisations • Decentralisation Act emphasis on devolved management of natural resources • Emergence of NGO voices – social and environmental activists • Government sanctioned handing over of all biotic resources to identified user groups • 100% of benefits flow to user groups • Land tenure remains with government • Government retains right to reclaim forests if misused by local people • Timber exports banned

wood-based industries, thus effectively moving community forestry away from a sole focus on fulfilling subsistence needs to a recognition that user groups can manage forests for commercial objectives (ibid.). Community forestry, or user group forestry as it is also known, has now entered a new era where much of the experimentation with new institutional formats has now been formalised and the more difficult task of support and development of new activities is now facing Forest Department staff. This is a period when the role of the public sector is coming under increasing scrutiny as local organisations take on more forest management roles. The implications for the public sector are considered in more detail in Chapter 7.

3.8 Discussion

Despite the major political and social differences between India and Nepal there is remarkable similarity in the evolution of forestry practice in both countries (see Table 3.3). There are, of course, significant differences in terms of the detail of implementation but an assessment of the broad impacts on forests over nearly 200 years leads to similar conclusions about the emergence of devolved forms of forest management, with many of the same contradictions between policy and practice. In many respects this should be no surprise, particularly with the globalisation of aid and thus the rapid spread of ideas. On the other hand, it is interesting to note how a nation (Nepal) that remained relatively isolated from the world for a hundred-year period was in fact deeply incorporated in terms of ideas and export of products – in particular, timber.

The significant differences that exist between India and Nepal lie in the legislative framework: in Nepal 100% of the forest resources are legally transferred as a right to the local people, whilst in India the rights to share the forest products are only granted administratively and are not a legal right. There are exceptions to this across the States, most notably Himachal Pradesh and Uttar Pradesh (the case of the **van panchayats**) (Campbell and Denholm, 1993).

Notes

1. This chapter draws on two PhD theses: Hobley, 1990 and Malla, 1992
2. Grant of land to individual nobles as a reward for some service rendered to the state. It was usually both tax free and heritable, and was valid until recalled or confiscated (Regmi, 1978)

4 Local Participation and Management Partnerships

> What is perplexing, as well as dangerous, is that scholars are willing to propose the imposition of sweeping institutional changes without a rigorous analysis of how different combinations of institutional arrangements work in practice ... Limiting institutional prescriptions to either 'the market' or 'the state' means that the social-scientific 'medicine-cabinet' contains only two nostrums. (Ostrom, 1994)

4.1 Introduction

This chapter looks at the array of institutional arrangements for the implementation of participatory forestry in India and Nepal. It considers some of the generic issues as well as the more specific questions that have arisen in each country. It looks at what constitutes a robust local organisation and the relationships that affect its

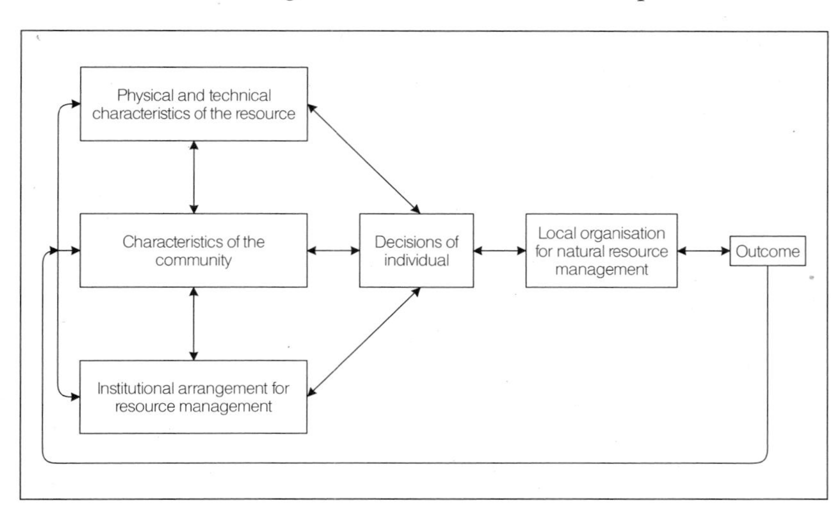

Figure 4.1
A framework for institutional analysis
(Scherr *et al.* 1995)

functioning (Figure 4.1). This leads us to a discussion in Chapter 5 of 'who benefits'. Appendices (A–F) provide discussion questions and exercises to illustrate some of the issues covered in this chapter.

Chapter 1 introduced the continuum of participatory processes and different levels of participation which may be appropriate to different contexts and outcomes. The skill lies in identifying when one form is more effective than another to achieve the objectives of a particular stakeholder group. With the advent of participation has come an additional dictionary of words to describe the individual's involvement in the participatory process. Hence, in participatory forestry projects there are users, interest groups, and stakeholders (also beneficiaries and target populations). The following sections give brief descriptions of the meaning of these words, and examples of how they are used, before considering the specific use of these words in India and Nepal.

Collecting oak leaves for animal fodder, Nepal

Collecting sal leaves for making plates, Nepal

4.2 Who Are the Local People?

The users

After 10 to 15 years of experience in Nepal where it appeared that participation was occurring at the passive end of the continuum, the concept of a forest user was introduced as being the defining feature for membership of a local forest management organisation. Following the development of the Operational Guidelines (HMGN, 1992) for the implementation of community forestry, users have been divided and categorised by their use of forests :

Primary user those who regularly use the forest area and have locally recognised rights to obtain all their forest product needs.

Secondary user those who occasionally use the forest area for a specific purpose or to obtain a specific product and are not given full rights by the primary users to obtain all forest products.

In addition to user groups, there are other groups within local and regional societies who have an interest in the forest. These are generally referred to as interest groups and can be defined as: 'people who have an interest, opinion or impact on a resource or area'. In most cases, these groups are easily identified through the products they collect and trade. For example, a major group in the hills of Nepal, who would not necessarily be easily identified as a user group, is the medicinal herb collectors. Management interventions impact on their use of forest products, but because of its dispersed and seasonal nature, they are rarely consulted and the impacts on them consequently often go unobserved and unmeasured.

What is a stakeholder?

Stakeholder is a term which, over the last few years, has come into common usage by most donor organisations; it was first used in business management theory and has since been widely adopted as a further refinement of the user concept. It is an umbrella term which covers all the people and organisations who have a stake in and may be *affected* by an activity, a development programme or a situation or who may have an *impact* or *influence* on it. In some situations stakeholders may both be affected by the intervention and also have an impact on the intervention. Box 4.1 provides an example of the stakeholder groups identified by the World Bank as being of importance in their programmes. As can be seen from this

Box 4.1 Stakeholders and the World Bank

The Bank's mandate and development objectives require it to recognise different stakeholders:

Borrowing Stakeholders: governments of borrowing member countries, including publicly owned entities and their staff.

Primary Stakeholders: those expected to benefit from or be adversely affected by Bank-supported operations, particularly the poor and marginalised.

Secondary Stakeholders: those with technical expertise and public interest in Bank-supported policies and programmes, as well as those with linkages to primary stakeholders. For example, NGOs, intermediary organisations, private sector businesses and technical and professional bodies.

Source: World Bank, 1994

example, the concept does have a wide interpretation, and its use indicates an overt acceptance of the role many individuals have in the identification and implementation of development projects. It signifies a new awareness of the importance of stakeholders' involvement in the project cycle and in policy construction.

In general stakeholders are classified in the following way (Grimble *et al.*, 1994; Reay, pers. comm.):

Primary Stakeholder describes people and plants and wildlife who are wholly dependent on the resource or area (e.g. forest area) for their survival. They often have few other livelihood choices or no choice in the short term and thus if a change comes they have difficulty in adapting. Geographically they usually live in or very near the resources in which they have a vital stake. There is often debate about whether forestry officers or people in forest industries are primary or secondary stakeholders. In most stakeholder analyses the term primary stakeholder refers to forest resource-dependent individuals, whereas government officials charged with the management and industrial users are considered to be secondary. There are obviously unclear areas where small-scale industrial users may be considered primary stakeholders.

Secondary Stakeholder includes all people and organisations who have a stake or interest in the resources or area under consideration – including industrial and governmental organisations.

Micro-level Stakeholders local and small-scale groups who are immediate users and real managers of the resources through their daily actions.

Macro-level Stakeholders regional and national planners, government departments (at the centre or state level), the global community, global consumers.

Stakeholder Analysis is the process of describing the nature of the stake which stakeholders have, or the characteristics and attributes of stakeholders. There are several continuums which can be used to distinguish between stakeholders and their stakes such as geographical location (proximity to the resource), time factors (people today, future generations) and power and dependency. Table 4.1 looks at the continuum of interest from the macro to micro level. Analysis of the table indicates the degree of divergence that can emerge between the interests of different groups.

4.3 What Is an Institution?

The word 'institution' or 'institutional arrangements' encompasses a broad set of meanings. There are two main complementary concepts which underpin this analysis of institutions: (i) regulatory arrangements such as customs or sets of rules, values or practices accepted by members of a particular group and which tend to lead to repetition of patterns of behaviour, and (ii) organisational arrangements which include ordered groups of people such as a family, farm, private firm, non-profit or governmental agency (Gibbs, 1986 quoted in Fox, 1991; Uphoff, 1986, 1992, 1993).

In forestry there are several important levels of interpretation of what constitutes an institution which will be discussed in this chapter, and in Chapter 7, in the light of the decentralisation policies operating in India and Nepal, namely:

Table 4.1 A continuum of stakeholders

Continuum level	Example stakeholders	Forest interest
Global and international society	International agencies, foreign governments, environmental lobbies, future generations	Biodiversity conservation, climatic regulation, empowerment, local rights
National	National governments, macro planners, urban pressure groups, NGOs	Timber extraction, tourism development, resource and catchment protection, empowerment, local rights, equity
Regional	Forest Departments, regional authorities, downstream communities	Forest productivity, water supply protection, soil depletion
Local off-site	Downstream communities, logging companies and sawmills, local officials	Protected water supply, access to timber supply and other forest products, conflict avoidance
Local on-site	Forest dwellers, forest-fringe farmers, livestock keepers, cottage industries, forest product collectors (for the market)	Land for cultivation, timber and non-timber forest products, cultural values

Source: adapted from Grimble *et al.* 1994

- property rights institutions
- formal institutions (government Forest Departments – covert and overt)
- non-formal institutions for resource management (extant or new)

There are several ways in which local organisations for forest management may emerge: as a local (indigenous) response to resource management; or as an externally catalysed response. We look at some of the indigenous organisations for forest management in both India and Nepal and consider some of the issues surrounding the interface between new and existing organisations. Chapter 7 looks at the importance of 'key individuals' in this process, and considers questions of institutional sustainability and the role of the catalyst through the experiences of NGOs, volunteer organisations, bilateral projects and government staff.

The property rights continuum

Underlying the move towards the decentralisation of resource control and management lies the assumption that it will lead to more efficient, equitable and sustainable resource use. The debate now centres on what type of institutional arrangement is most appropriate in a given social institutional context. Aspects of these arrangements include property rights structures as well as organisational structures.

At one end of the property rights debate are those who state that total privatisation of resources to rational individuals will lead to more efficient and sustainable use (Demsetz, 1967), whereas at the other end of the spectrum the common property[1] literature points to the potential of sustainable group management of forests, where there are adequate individual incentives, secure long-term tenure arrangements (Fortmann and Bruce, 1988) and group-imposed restrictions.[2] Ostrom *et al.* (1988) detail many cases that indicate that there are situations in which co-operation between a group of resource users does lead to careful and sustained management. The work of Netting (1976) in Switzerland and McKean (1986) in Japan provides further evidence to support the effectiveness of collective management under certain conditions,[3] as does the large and detailed literature from Nepal. (For a good analytical discussion of this literature see Fisher, 1989; 1991). McKean (1995) questions the form in which property rights should be divested and poses some important questions, aspects of which are addressed in this chapter and in Chapter 5:

- In whom (to how many persons, to which persons, with what distributional consequences) should property rights be vested?
- Which rights should be transferred – full ownership with rights of transfer, or just use rights?
- What kinds of resources should be privatised? Are all objects equally able to be divided up? Should ecosystem boundaries matter?

Others, most famously Hardin (1968), have contested the assertion that local people can be effective resource managers and argued for highly centralised structures in order to protect the ecological integrity of a resource and avert a 'tragedy of the commons'. However, in his paper Hardin fails to distinguish between resources that do not have a property regime i.e. open access, and those which do (Hobley, 1985). In a recent paper, he has attempted to rectify this confusion and distinguishes between managed and unmanaged resources (Hardin, 1994). The earlier view promoted by Hardin coincides with the views commonly heard, that the peasants are the destroyers of the environment whereas the government is the custodian. Joint forest management challenges the central tenet of this argument, and posits the view that under certain circumstances local people, together with the state, should become the managers of the forests. Community forestry in Nepal moves a step further forward and asserts that the state's role is that of regulatory authority only and that total management control should rest with the users of the resource (though, property rights are retained by the state). Under these rulings there is a clear understanding that the state can no longer take sole responsibility for the management of forests – since organisationally it does not have the capacity to ensure the integrity of the resource into the future without the co-operation of those who use the forests.

The usual dichotomy drawn between public and private management can only be considered helpful in the early stages of analysing institutional options for a particular sector. The continuum approach, however, provides the most interesting way forward and perhaps the most pragmatic. Runge's (1986) analysis reinforces the observation that there are no simple property rights scenarios, but rather that there is a continuum of options that need to be put in place according to the particular conditions and context of the resource:

> rather than invoking the general superiority of one type of property institution, . . . different institutions are responses to differing local environments in which institutional innovation takes place. Such innovations are likely to range along a continuum of property rights, from pure rights of exclusion to pure rights of inclusion, depending on the nature of the resource management problems . . . There are not universal prescriptions for efficient and equitable resource management.

The collective end of the continuum

As Ostrom (1994) so cogently argues, it is not an either/or situation but an 'and' situation where there are many arrangements that can be accommodated ranging from partnerships between government and local people to complete local control. Using this notion of a continuum there are a variety of institutional arrangements that could be selected according to the particular context. This approach requires site specificity and a high degree of social contextual understanding from the implementing or facilitating organisation. To date, although appealing to academics in its recognition of complexity and diversity, it has been resisted by government institutions used to the prescriptive model-based approach to development. This is discussed further in Chapter 7.

In the case of forests in both India and Nepal where the land on which they are growing is clearly vested in the government, the association of institutions is clearly defined by this central tenet. The decision about the form of institutional partnerships therefore revolves around the extent to which Forest Departments should retain authority over management decisions for an area of forest and over usufructuary rights, though there is little debate as to whether the government should or should not retain control over the land. Indeed, joint forest management is seen by some within Forest Departments as a means to reassert control over forest lands and defend their boundaries, and by others as a fundamental challenge to their authority. Thus the choice for collective management is described by Figure 4.2.

The question to be addressed is: what are the conditions neces-

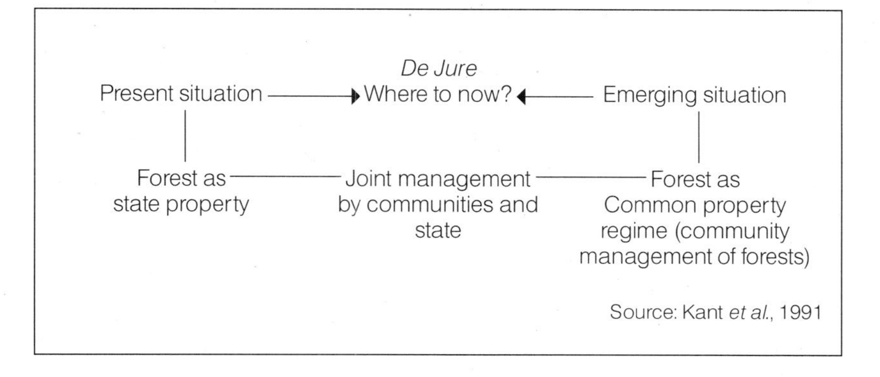

Figure 4.2
The forest management continuum

sary to trigger local people to implement their own institutional arrangements to change the structure of the situation in which they find themselves? (Ostrom *et al.*, 1988) The answer is complex. One of the key enabling structures is the presence of a facilitating policy framework; for example, community forestry in Nepal gained greatest impetus once government had passed the 1993 Act which then allowed guidelines to be written that provided Forest Department staff and users with legitimacy for their actions. This policy, most particularly, affirmed the legitimacy of local people's usufruct rights. Similarly in India, prior to the passing of the JFM resolution very few new partnerships for forest management emerged, except in those cases of enlightened and innovative action by local government, NGO staff or local leaders. Thus the importance of changing the property rights institutions is fundamental to the development of any form of participatory forestry:

> (These) critically affect incentives for decision-making regarding resource use and hence economic behaviour and performance. By allocating decision-making authority, property rights also determine who are the economic actors in a system and define the distribution of wealth in a society. (Libecap, 1989)

Under what conditions is privatisation the answer?

There are a series of calculations that need to be made before the decision to privatise can be taken. Much of the literature concerning the medieval open field systems (Dahlman, 1980) addressed this question and showed that privatisation of resources was only possible when the costs of protecting the individual's boundaries did not outweigh the benefits of production. In the case of forests, except for small patches of forest close to a villager's house, it is virtually impossible to protect the forest against the predations of outsiders. In such circumstances the costs of individual protection would far outweigh any benefits; it therefore makes sense for a group of forest users to come together to manage the resource in common, thus spreading the costs of protection across a larger group of people (McKean, 1995). In some senses, this could be considered to be group privatisation, where a group of individuals are 'a private owner that share property rights for managing common pool resources' (McKean, 1995; McKean and Ostrom, 1995).

The utility of such an approach depends on the extent of benefit obtainable: if the resource is of sufficient extent or value (not necessarily financial) there is sufficient reason for individuals to manage it in common, again with the proviso that there is security of tenure over the resource that can be upheld both against the power of the state and against locally powerful non-rightholders. Thus, common management with its attendant rules and punishments for infringement demands a degree of individual responsibility to each other's neighbour, and does not permit the individual to ignore the effect of his or her actions on others. As has been demonstrated by McKean (1995), it is likely that as populations increase common property regimes will become more rather than less desirable in those areas where 'prevailing cultural values support co-operation as a conflict-solving device'.

'The transfer of property rights from traditional user groups to others (private individual or public ownership) eliminates incentives for monitoring and restrained use, converts owner-protectors into poachers, and thus exacerbates the resource depletion that it was supposedly intended to prevent'.

(McKean, 1995)

In forestry, the task facing forest authorities is to identify the institutional responses relevant to a particular social, ecological and political context (Leslie, 1987). In some situations collective action will be an appropriate response (Table 4.2); in others it may not be possible to evoke a collective response. Such situations may require other responses such as small group leaseholds, or perhaps privatisation of resources (in most cases to groups, but in some instances to individuals or to private sector companies). All change will require the provision of an enabling policy framework, and assured long-term rights of access.

4.4 Local Institutions

There are few aspects of local institutions which are able to be generalised: an assessment of forestry user groups in Nepal (Chhetri and Pandey, 1992; Karki *et al.*, 1994; Dahal, 1994) and joint forest management groups in India (Kant *et al.*, 1991; Sarin, 1993) indicates a great degree of variability in rules, use-rights, etc. As Ostrom (1994) states, however, it is this variability that provides some pointers to institutional sustainability. One would question the appropriateness of institutions that all had the same use rules irrespective of ecological or social variation. 'By differing, the rules, take into account specific attributes of the physical systems, cultural views of the world, and the economic and political relationships that exist in the setting.' Ostrom outlines seven design principles that characterise most of the robust common-pool resource institutions. These are amalgamated here with a list produced by Wade (1988) and presented in Box 4.2. They are used to assess the robustness of some of the local organisations described in the case-study material presented in this chapter.

Table 4.2 The continuum of joint practice for different forest management objectives

Objectives		
Conservation	**Economic**	**Local**
Users 1. Identification of all users of the forest resource, e.g. seasonal plant collectors	Identification of all users of the forest resource and potential other stakeholders in the management process (industry through to local people)	Identification of all users of a forest area. To include all villages surrounding the forest (whether they have formal rights or not) and seasonal users
Consultation 2. Joint consultation and planning with users, Forest Department, conservation groups, to determine how best to manage biodiversity without removing users' access to livelihoods	Joint planning of the forest area to determine specific management objectives and means to achieve them This stage will also require detailing of responsibilities and controls for extraction and mechanisms for policing extraction process	Development of consensus about how forest is to be managed and who is to be allowed to use resource, under what conditions; agreement about usage boundaries, etc.
Plans 3. Production of joint plan agreed by all users and implemented by identified organisations (Forest Department with neighbouring villages, advice from conservation groups)	Production of agreed management plans (Forest Department primarily with private sector and local group, where organised)	Joint production of a forest management plan to be implemented by local people Joint planning of future forest land management strategies (particularly in adjoining areas of economic or conservation forestry) Liaison with other JFM groups in area, exchange of experience, joint planning of protection and management. Possible future joint marketing systems.
Benefits 4. Benefit-sharing and compensation packages. Sliding scale of financial benefits dependent on degree of involvement in management	Benefit-sharing in these forests initially would be as employment, later possibly through contract harvesting arrangements to local groups	Benefit-sharing dependent on degree of management, protection activities. Would expect to move towards a sliding scale approach (including a base-line assessment of the market value of the forest)

Box 4.2 How Robust are Local Forest Management and Protection Institutions: Criteria for Assessment

User group

- Size: the smaller the number of users, the better the chances of success, down to a minimum below which the tasks able to be performed by such a small group cease to be meaningful. Swallow and Bromley (1994) suggest, from their research, that a group agreement is more likely to collapse where there are more than 30–40 members.
- Boundaries: the more clearly defined are the boundaries of the group, the better the chances of success. Individuals or households with rights to withdraw resource units from the common-pool resource are clearly defined and agreed.
- Relative power of sub-groups: the more powerful are those who benefit from retaining the commons, and the weaker are those who favour sub-group enclosure or private property, the better the chances of success.
- Existing arrangements for discussion of common problems: the better developed such arrangements are among the users, the greater the chances of success.
- Extent to which users are bound by mutual obligations: the more concerned people are about their social reputation, the better the chances of success (Runge, 1986).
- Punishments against rule-breaking: the more the users already have joint rules for purposes other than common-pool resource use, and the more bite behind these rules, the better the chances of success.
- Consensus about who are the users: recognition of customary user rights as well as legal user rights is important. This must be negotiated at the outset of the formation of a collective action group.
- Distribution of decision-making rights and use rights to co-owners of the resource need not be egalitarian but must be considered fair (McKean, 1995).

Relationship between resources and user group

- Location: the greater the overlap between the location of the common-pool resources and the residence of the users, the greater the chances of success.
- Users' demands: the greater the demands (up to a limit) and the more vital the resource for survival, the greater the chances of success.
- Users' knowledge: the better their knowledge of sustainable yields, the greater the chances of success.

Box 4.2 (continued)

The technology

The higher the costs of exclusion technology (such as fencing, rotational patrolling), the better the chances of success, i.e. investment in the resource leads to a greater incentive to protect.

Congruence between appropriation (use) and provision rules and local conditions

Appropriation rules restricting time, place, technology, or quantity of resource units are related to local conditions and to provision rules requiring labour, materials, and/or money. There should be inbuilt flexibility in the use rules to respond to changes in the resource or to the economic environment (McKean, 1995).

Detection and graduated sanctions

- Users who violate operational rules are likely to receive graduated sanctions (depending on the seriousness and context of the offence) from other users, from officials accountable to these users, or from both.

- Ease of detection of rule-breaking free riders: the more noticeable is cheating on agreements the better the chances of success.

Collective-choice arrangements

Most individuals affected by operational rules can participate in modifying them.

Monitoring

Monitors, who actively audit common-pool resource conditions and user behaviour, are accountable to the users and may be the users themselves (see Chapter 7 for use of process documentation as a monitoring tool).

Relationship between users and the state

Ability of the state to penetrate to rural localities, and state tolerance of locally based authorities: the less the state can, or wishes to, undermine locally based authorities, and the less the state can enforce private property rights effectively, the better the chances of success.

Conflict-resolution mechanisms

Users and their officials have rapid access to low-cost local arenas to resolve conflict among users or between users and officials.

Nested enterprises

Appropriation, provision, monitoring, enforcement, conflict resolution, and governance activities are organised in multiple layers of nested enterprises (possibly linked with other democratically based political institutions).

4.5 Operational Experience

The following sections briefly review the actual implementational practices of community forestry and joint forest management, and subsequently consider the empirical evidence to date in order to assess the effectiveness of the development of forest management groups.

In Nepal, community forestry practice is guided by a set of comprehensive guidelines issued by the government and formalised in 1992. The guidelines define a four-phase planning process (Bartlett, 1992):

- Investigation phase — involving rapport-building with villagers, information-gathering and identification of the users and existing management systems
- Negotiation phase — involving consensus-building, formation of the forest user group, negotiation of a forest management agreement and writing of an operational plan (facilitated by Forest Department staff)
- Implementation phase — involving conduct of forest management, according to the operational plan, by the forest user group, with monitoring, support and strengthening activities conducted by the field staff
- Review phase — involving appraisal and renegotiation of the management agreements

Formation of a management system for a community forest involves a large amount of work: identifying different interest groups; discussing in small and large groups; verifying decisions made in group assemblies by conducting informal surveys amongst the users . . . The overriding premise behind all the work was that the users should decide for themselves the composition of the forest user group and the management arrangements for their forest, with the field staff acting as facilitators and providing technical advice where necessary.
(Bartlett *et al.*, 1992)

Clearly within these four phases there is a large amount of flexibility to allow field staff to progress and implement according to the particular situation. Problems arise when there is rigid adherence to a series of steps without the necessary responsiveness to local conditions. This latter problem is more apparent in India, where there has been a more mechanistic and prescriptive implementation of joint forest management. This is perhaps a reflection of the earlier development of approaches compared with Nepal, where initially some of these problems of prescriptive implementation were more apparent (Nurse and Chhetri, 1992). Recent development of manuals and guidelines in different States has attempted to develop guidance for more responsive approaches to local conditions.

After a somewhat slow start, the community forestry programme and the formation of forest user groups in Nepal are now proceeding at an accelerating rate. Table 4.3 indicates the current status of community forestry through forest user groups in four districts in eastern Nepal under the Nepal-UK Community Forestry Project. It is evident that community forestry is now beginning to involve significant numbers of people, and considerable areas of forest.

In India too, JFM is expanding rapidly following a phased approach similar to that described for Nepal; some would say too

Table 4.3 Current status of forest user group establishment in 4 hill districts

District	No. of user groups	No. of households involved	Total community forest area (ha)	% of total district forest area handed over	% of district population involved
Bhojpur	122	11,500	6,614	16	31
Dhankuta	125	11,786	6,968	48	43
Sankhuwasabha	71	5,354	3,901	11	20
Terhathum	120	9,441	4,904	32	51
Total	438	38,081	22,387	av. 27	av. 36

av. = average

Source: NUKCFP forest user group database, September 1994, and Stewart, 1987.

quickly in States such as Orissa, where over 6,000 committees were registered over several months, and now relatively few are functioning. The question of speed versus quality is one that still remains to be answered, as empirical evidence for what constitutes a 'good' local organisation is still being collected and analysed.

4.6 The Search for Appropriate Local Organisations

In both India and Nepal considerable debate has centred around the question of what is the most appropriate institutional structure for participatory forestry at the local level. It revolves around two issues: the recognition and role of informal, indigenous local management systems versus the imposition of more formal externally developed organisations (Gilmour and Fisher, 1991).

With the shift in focus towards participatory forestry in both India and Nepal foresters and researchers have begun to identify existing indigenous management systems in a wide diversity of social and ecological systems (Acharya, 1989; Arnold and Campbell, 1986; Campbell *et al.*, 1987; Baral, 1991; Baral and Lamsal, 1991; Bartlett and Malla, 1992; Chhetri and Pandey, 1992; Fisher, 1989, 1991; Furer-Haimendorf, 1964; Gilmour and Fisher, 1991; Jackson, 1990; Jackson *et al.*, 1993; Kant *et al.*, 1991; Karki *et al.*, 1994; Mahat, 1985; Molnar, 1981; Messerschmidt, 1984, 1986, 1987; Raju *et al.*, 1993; Sarin, 1993; Tamang, 1990). In Nepal, forest management systems may include those under government-sanctioned systems during the Rana rule, such as the **kipat** and **talukdari** systems in which the government maintained a right of control and taxation, religious forests, and systems indigenously derived by villages, households or clans. A number of studies reveal that these systems have been in operation for decades, while others appear to be recent responses to a changing institutional framework. Many of them appear to be robust in terms of their

Box 4.3 The Form of Forest Management and Institutional Arrangement

1. Government forests Examples: Reserve forests, high value protected forests, and areas of conservation importance (Nepal and India)	• Forest Departments retain the authority and responsibility for control and management • Objectives of management are protection and regulation of timber and other products with a cash value; issuing of licences to harvest forest products • Local people considered to be a danger to the forest resource • In the absence of consistent administration and enforcement, these forests have become virtually open access resources
2. Indigenous Forests Examples: **shamlat** forests (Jammu & Kashmir, Punjab) sacred groves (India and Nepal)	• Forests which are legally owned by the government but which have been managed by local people on their own initiative • Forest Departments consider these to be government forests, but the local people, who have protected and used the forests, consider them to be their own • Local people are *de facto* managers • Such systems usually associated with patches of natural forest • Usually have clearly defined boundaries and users with use rights • Tend to have a strong leader • Use consensus and sanctions to protect and manage forests • Employ own watcher or use rotational watching systems • Highly conservative, protection-oriented systems
3. Externally imposed or sponsored forest management: Examples: Community forestry, Hill Resource Management Societies, forest protection committees, (under some JFM initiatives)	• Forests which are legally owned by the government. The impetus for development and management of common forest resource is derived from outside the community • In most cases, these forests were developed and managed without recognition of the existing indigenous management system • All physical and financial inputs are supplied, and decisions tend to be taken externally • Forests managed by outside agency through appointment of watchers • Oriented towards resource creation and protection, with emphasis on achievement of technical and not socio-economic objectives • Assumes responsibility should be given to local elected leaders, as the representatives of local opinion • Authority and control remain with the outside agency and not the local people • Infringement is usually punished externally and not through internal mechanisms • All harvesting decisions are made by the Forest Department • Systems characterised by mistrust between local people and Forest Department staff • Local people are passive and not active participants • Hands-on management by Forest Department staff cannot be sustained across hundreds of patches of forest • Into this category fall all the forests developed under early social and community forestry project-based approaches, and past systems such as the **talukdari** system • Under this characterisation also fall some forms of current community forestry and joint forest management practice

Box 4.3 (continued)

4. Joint Forest Management An ideal scenario	• Active partnership between forest users and outside agencies • Draws on knowledge, experience and expertise of both outsiders and local people to manage forests • Focuses on the user groups rather than the politico-administrative unit as the unit for management • Development of self-sustaining groups capable of articulating needs and sourcing services
The real scenario Examples: user group forestry (Nepal); **gram vikas mandal** (Gujarat); forest protection committees (under JFM arrangements)	• Authority and control still vested with the Forest Department • Ignores differential interests within user groups • Forest Department staff still directing most management operations

management ability and maintenance of access to productive forests (Bartlett and Malla, 1992; Chhetri and Pandey, 1992; Karki *et al.*, 1994).

For Nepal, Malla (1992) has identified four broad categories of forest that can be categorised on the basis of where control and authority lie. These have been amalgamated with a similar classification for India made by Sarin (1993) and are described in Box 4.3. Management ranges from extensive systems limited to protection and some harvesting, to intensive management using different silvicultural techniques.

As can be seen from this categorisation, these are not four discrete forms of forest management; rather they all co-exist, with some forms moving forward into new arrangements and others regressing into previous forms of control.

India also possesses a diversity of historically recognised local management systems including the Forest Co-operatives of Kangra District in Himachal Pradesh, the **van panchayats** of Uttar Pradesh, and the Communidades of Goa (Tucker, 1982; Chambers *et al.*, 1989; Gadgil and Guha, 1992b; Singh, 1992; Agrawal, 1994; Britt-Kapoor, 1994). What has been remarkable (and poses some interesting questions about the invisibility of informal systems to outsiders) has been the discovery and documentation by a number of researchers of widespread indigenous forest protection and management systems in tribal areas of Orissa, Bihar, Karnataka and Gujarat (Kant *et al.*, 1991; Raju *et al.*, 1993; Sarin, 1993; Sarin and SARTHI, 1994). No doubt there are other cases in other States, but as yet the appropriate questions have not been asked. In Orissa and Bihar alone several thousand indigenous forest management groups have been identified, protecting over 200,000 ha of forest land (Singh and Singh, 1993). Each has different characteristics, membership criteria,

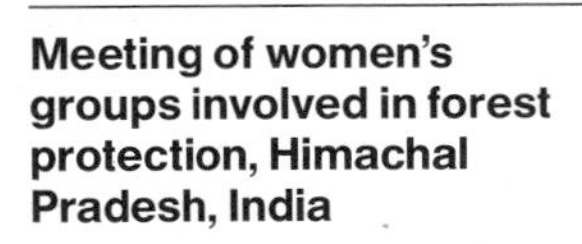

Meeting of women's groups involved in forest protection, Himachal Pradesh, India

rules, regulations, sharing arrangements and degrees of formal structures.

The Indian situation is further complicated by a diversity of local forestry organisations initiated and facilitated by NGOs. Inevitably the structure of these groups reflects the development philosophy of the facilitating organisation, from the AKRSP approach of multi-functional organisations to the single-focus forestry approach of VIKSAT and other NGOs. In addition, women's groups (Mahila Mandals) and youth groups (Yuvak Mandals) have, in Himachal Pradesh, carried out planting and protection work on forest lands. Such diversity provides a rich tapestry from which to analyse the particular conditions under which some organisational structures may be more effective than others.

4.7 Analysis of Experience with Resource Management Organisations

The following sections begin to assess these criteria, accepting that there is a large amount of work still to be done to ensure that the rigour of the criteria can be demonstrated. The evidence draws on experiences from indigenous as well as outsider-initiated groups.

Location and clearly defined boundaries

In most of the Middle Hills areas in Nepal there are clearly identifiable users based on residence and proximity to the resource to be managed. Since the forest patches are relatively small, it is also relatively easy to identify the boundaries of the resource, and in many cases these have already been negotiated on an informal

Small areas of forest make forest protection easier, Nepal

basis. However, the previous practice of allocation of forests to the panchayats and not to the users did mean that some committees set up under that system excluded many of the traditional forest users whose residence happened to fall outside the panchayat boundary, leading in many cases to non-functional committees and disenfranchised users (see King *et al.*, 1990). As discussed in Chapter 3, this has since been resolved where legislation now recognises the primacy of the user and is not based on a relatively arbitrary politico-administrative boundary.

However, there are many areas in Nepal where forest ownership is still contested, following poor survey and demarcation under the cadastral survey which has taken place across Nepal over a 60-year period. This has led to many cases of conflict when user group forestry has been introduced into these unresolved land-tenure situations (Baral, 1993; Kafle, n.d.). There are similar experiences in India where serious disputes, ending at best in destruction of the forest and in the worst case in death, have occurred over boundaries and access rights following the introduction of forest protection systems (Raju *et al.*, 1993; Sarin, 1993; Saxena, 1992).

Just as in Nepal where policy and practice have moved away from handing management to the lowest administrative unit, so experience in India is now revealing that a reliance on the formal structures to represent forest users is also not effective. Rather than relying on legal and administrative boundaries to provide the criteria for assessing whether an individual is a user or not, it is necessary firstly to identify who has the right to be a member of a local forest organisation. In the case of indigenous organisations membership is based on who has negotiated customary rights to a forest. Hence in many cases customary rights are exercised outside the revenue boundaries of one village in an entirely different revenue village (see Box 4.4).

Problems still arise in areas of extensive forests where there are multiple users, such as are found in high altitude forests and in the last areas of tropical moist forests in India (Box 4.5). Here a complex of users differing by season and product, may have an interest in the management of the forest. In some cases user groups have

Box 4.4 Boundaries and Control

Among the more complex aspects of the self-initiated forest protection groups (SIFPGs) in Gujarat are the processes of inter-group negotiations by which composition of the social unit of organisation and the boundaries of the forest area it has started protecting, have been defined. Because of the scattered settlement pattern of the area's tribal and semi-tribal population, most villages near the forests do not have compact settlements. Instead, they consist of individual houses scattered next to agricultural landholdings. Most villages have a number of **falias** (hamlets) named after the particular sub-caste or tribe residing there. However, even these are not easily identifiable as physical units, since essentially they are *social* units. Administrative 'revenue village' or 'gram panchayat' boundaries do not necessarily overlap with the social unit boundaries. There is also considerable variation in the amount of forest land within the boundaries of different revenue villages, with little correlation between a village's population and the forest area within its boundaries.

Yet the majority of the population continues to have similar levels of dependence on forest produce. As a consequence, it is physical , extent of dependence, and social relations, rather than formal 'revenue village' boundaries, that have determined which people use which forest area. It is the former, rather than the formal boundaries, which have determined the membership composition of the SIFPGs.

Source: Sarin and SARTHI, 1994

Box 4.5 Strategies for Resolving Access Conflicts Between Primary and Secondary Users

Due to certain user groups living at distances of 3 to 5 km from the forest, and therefore being unable to participate in day-to-day protection and management, a 'two-tiered' access structure is beginning to evolve in Haryana. While the primary users, who tend to reside nearer the forest area, enter into a joint management agreement with the FD and accept primary management responsibility, they also agree to permit continued access to the non-member secondary users on clearly defined terms.

Similar two-tier access structures have also been evolved by many of SIFPGs in Orissa. In Kishorenagar block, the residents of 17 or 18 gram panchayats have started protecting forests near their villages. They permit villagers of three or four other panchayats which do not have forests in their vicinity to collect forest produce from the protected forests. Similarly, the village forest protection committee (VFPC) of Budhikhamari in Mayurbhanj district of Orissa, the formation of which was facilitated by an officer of the Orissa Forest Department prior to the state government resolution being issued, permits its own members to collect sal (*Shorea robusta*) leaves free of cost, whereas non-members are allowed access for a small fee of Rs 2 per person.

Source: Sarin, 1993

resolved this dilemma by forbidding access to the forests by users who are not resident in the area (Karki *et al.*, 1994). In others, a sliding scale of access to benefits according to residence status has been used: from 100% benefits for long-established households to 50% of the benefits for temporary residents (ibid.). In other cases, particularly where nomadic pastoralist groups are involved, there are few examples of successful conflict resolution between the settled users and the nomads (Eagle, 1992; Sarin, 1993; Vira, 1993). Potentially this is an area of major conflict, since many settled villagers have stated that they are attracted to the joint forest management approach as it allows them to control access to their forests, and in particular to forbid pastoralist use of the forests. The impacts on different user groups are discussed further in Chapter 5.

Users' demands

The relationship between resource scarcity and collective action appears to be relatively straightforward (see Gilmour, 1990; Gilmour and Fisher, 1991; Arnold and Dewees, 1995; Thomas-Slayter, 1994 for a description of these relationships and Table 4.4; also Kant *et al.*, 1991 for a discussion of local management systems in Orissa). For example, in some areas of Nepal there are well-established systems of forest management in areas of previously scarce resource (as was described in Chapter 3). Equally, however, there are examples of collective action in areas of high resource availability (Budhathoki, 1987). Thus, it is not simply scarcity that drives local initiative; it also requires leadership, consensus on

Ecologically marginal areas require significant investment before there are noticeable gains to production, Rajasthan, India

action to be taken, ability to enforce restrictions and confirmation from government that panchayats (or other local organisational units) are empowered to take such action. Thus, although the ecological scarcity to local action equation looks to be a simple causative link, and is a useful broad-scale planning tool, many other factors need to be considered (Chambers *et al.*, 1989; Fisher, 1991; Loughhead *et al.*, 1994; Arnold and Dewees, 1995).

There are also conditions where the land is too degraded or ecologically fragile, and thus the investment (both financial and human) to bring it into production is too great for local people to undertake. In such cases, it is unlikely that collective action will be possible since the future benefits are uncertain and the immediate costs very high. Thus, there are broad categories that can be assigned to resources which equate to the type of management partnership likely to be effective. For example, in highly degraded areas, it is likely that the government will have to take the major role in their regeneration (see Chapter 5 and Table 5.1 for further discussion of this issue).

In addition to the extent of the public resource, a further important factor is the degree of individual access to private tree resources. It has been found in numerous studies that, as public resources decline, the individual begins to invest in planting and protecting trees on private land (Box 4.5; Carter and Gilmour, 1989; Chambers *et al.*, 1989; Carter, 1992; Hobley, 1990; Malla, 1992; Arnold and Dewees, 1995). In addition, intra-household factors, such as availability of labour, also have an important impact on whether trees are planted or not (Dewees, 1995). Table 4.5 provides a useful comparative summation of the factors influencing the planting of trees on private lands by small farmers.

However, as a number of studies have shown, tree planting is a

Table 4.4 From plenty to scarcity

Resource	Local interest	Response
1. Ample forest in or adjacent to village	No interest in forest protection or tree planting	Indigenous management systems exist, confined to defining use rights only. Few trees on private land
2. Forest becoming depleted or access restricted (up to 3-hour walk)	Emerging interest in forest development activities	Indigenous management systems exist to define use rights and in some cases have biological objectives. Some trees on private land
3. Severe shortage of forest products (accessible forest more than 4-hour walk)	Genuine interest in forest development activities. Little need for people to be pursuaded.	Indigenous management systems well-developed and define both use rights and biological objectives. Extensive private tree planting and protection

Source: Gilmour and Fisher, 1991

Table 4.5 Why small farmers do and do not plant and protect trees

Factors	Do not plant and protect	Do plant and protect
Land tenure	• Insecure	• Secure
Rights to usufruct	• Shared with government, or subject to taxation or control, or ambiguous	• Vested entirely in the household, regularly exercised without restriction or rent
Choice of species	• To meet officials' priorities	• To meet farmers' priorities
Ownership of trees	• Owned by or shared with government or local authority, or ambiguous	• Owned by the household in law and in practice
Protection	• Hard to protect	• Easy to protect
Rights to cut and fell	• Restricted or believed to be restricted	• Unrestricted and practised at will by owners
Rights of transit to markets	• Restricted or believed to be restricted	• Unrestricted and practised at will by owners and buyers
Marketing	• Monopolistic	• Competitive
Market prices	• Not known, unstable or dropping	• Known, stable or rising

Source: Chambers *et al.*, 1989

Box 4.6 Tree Planting in Response to Livelihood Security

Tree planting can be explained as being a response to four categories of dynamic change:

- to maintain supplies of tree products as production from off-farm tree stocks declines, due to deforestation or loss of access;
- to meet growing demands for tree products as populations grow, new uses for tree outputs emerge, or external markets develop;
- to help maintain agricultural productivity in the face of declining soil quality or increasing damage from exposure to sun, wind or water runoff;
- to contribute to risk reduction and risk management in the face of needs to secure rights of tenure and use, to even out peaks and troughs in seasonal flows of produce and income and in seasonal demands on labour, or to provide a reserve of biomass products and capital available for use as a buffer in times of stress or emergency.

Source: Arnold, 1995

viable option only for those households with adequate areas of land; poorer households will continue to rely on a degrading resource and in the absence of local management systems will be forced to travel longer distances to more productive forests (Dewees and Saxena, 1995 and Box 4.6). Equally, in the first years of local management when activities tend to be protection-oriented, these poorer households are also forced to travel elsewhere to non-protected forests (Kant *et al.*,1991; Loughhead *et al.*, 1994). As local forests begin to upgrade and to supply a flow of products, in theory those with inadequate private resources will be able to use the local forests (Karki *et al.*, 1994). Those with adequate access will have little interest in becoming involved in management of the public resource, thus reducing the number of households deriving benefit from a limited resource. However, as is discussed in Chapter 5, this simple relationship is rarely demonstrated, although Karki *et al.* (1994) do cite one case where villagers with adequate private resources have waived their rights to the user group forest.

This brings us to another interdependent factor: the relative power of sub-groups. Many studies have indicated that this is a very important factor in determining the impact of forest management on different groups. There are cases where marginalised groups whose livelihoods depend on the forest have little involvement in decision-making and have been denied access to the resource under new stringent protection rules (Hobley, 1990; Sarin, 1993). These impacts are discussed in greater detail in Chapter 5.

There are also questions of inter-group equity where one user group may have more than adequate access to a forest, and another, perhaps adjoining, village has extremely poor access. Resolutions of these conflicts and urban-rural redistribution systems are going to become major factors affecting the success of participatory forestry (Kant *et al.*, 1991; Dahal, 1994).

Representation of different sub-groups on forest protection committees and decision-making fora is also an important factor in

Seasonal collection of leaves to make leaf plates provides an additional income source for poor households in the Tarai, Nepal

assessing of the robustness of the management system. Chhetri and Pandey (1992) found that where a decision was taken by one caste, it was ignored by other castes unless they too had been involved in reaching that decision (see also Sarin, 1993).

Some of the most successful local management initiatives have occurred where there are immediate benefits obtainable for the local groups (for example, the case of **bhabbar** grass in Haryana) (Arora *et al.*, 1993; Chatterji and Gulati, n.d.; Varalakshmi *et al.*, 1993).

Conversely, where local people have had to wait several years before there are any returns, interest has often declined and the resource in consequence has degraded due to the cessation of protection (Nurse and Chhetri, 1992). Greater success has also been achieved in cases where the products have a ready market and therefore there is perceptible value-added to the labour involved in protecting the resource (Malla, 1992). In recent years much more emphasis has been placed on the development of marketing structures for non-wood forest products (NWFP) and providing silivicultural systems that will optimise their production as opposed to the conventional timber-based systems (Kant *et al.*, 1991; Poffenberger and Sarin, n.d.; Edwards, 1995; Raju *et al.*, 1993).

With changing economic circumstances, the demand for wood and non-wood-based products (such as sal leaves for plate-making, silk worms to make tassar silk, and medicinal herbs) is also increasing, providing further support to the need to move community forestry and joint forest management approaches beyond subsistence to the supply of surplus products to these industries (Dutta and Adhikari, 1991; Malhotra *et al.*, 1991; Malla, 1992; Pachauri, n.d.; Sarin, 1993; SPWD, 1992; Varalakshmi *et al.*, 1993; Vijh and Arora, 1993). However, as was noted with farm forestry, markets are difficult to predict and products that may have a high value today may equally have a low value tomorrow. Thus in West Bengal, the sal leaf plate makers face a declining and highly competitive market; any management decision to maximise production of sal leaves should be considered in the light of these market constraints (Femconsult, 1995b).

Once a communal forest is managed to respond to market demands, it is more likely that powerful local interests will capture the benefits of production as opposed to small land holders and weaker members of the user group.
(Malla, 1993)

Recent work in both Nepal and India has been focused on trying to provide reliable assessments of the productivity of the forest and its ability to meet the demands of the users (Tamrakar, 1993; Malhotra *et al.*, n.d.). It became apparent from a review of operational plans that in most cases little attention had been given to these critical relationships, with the net result that user demands would probably exceed the supply, forcing users to exploit other unprotected resources (Branney and Dev, 1993; Loughhead *et al.*, 1994; Young, 1994). The methods used for these assessments are discussed in Carter (1996), and some of the silvicultural implications of meeting user demands are discussed more fully in Chapter 6.

Users' knowledge

For many years the role of the peasant in degrading the forest was an accepted dogma. In recent years this has been challenged and evidence has been gathered to indicate that there have been careful tree management systems in place, sometimes for several decades. Studies in Nepal and India indicate the importance and complexity of indigenous knowledge (Carter, 1992; Loughhead *et al.*, 1994). Some cases show how villagers have actively managed forests for

> 'One of the most valuable things I can do is to impress on the village people what a wealth of useful knowledge and skills they have. This knowledge of the uses of every kind of plant, the ability to survive with so little and making the most of what they need from what is locally available, is not low status knowledge'
>
> (VSO volunteer report)

particular products, for example, the removal of thorny species to facilitate the regeneration of oak discussed in Chhetri and Pandey (1992).

Group size and constituency

There is a certain logic to limiting the size of a group where decision-making is based on consensus. Intuitively, it would appear that small groups should be more effective than larger groups and this has been demonstrated to be the case (Chhetri and Nurse, 1992; Karki *et al.*, 1994; Malhotra *et al.*, n.d.; Moench, 1990; Poffenberger and Singh, 1992; SPWD, 1992). However, the empirical evidence to support this supposition is not clear; for example one study in Nepal indicated that large groups of over 300 households were not necessarily less effective than small groups of less than 100 households (Bartlett *et al.*, 1992). It appears that the interplay of various factors is more important than one single criterion. Hence, a highly factionalised but well represented and managed large group may be more effective than a non-factionalised but non-representative, poorly managed small group.

Some studies have indicated that ethnic homogeneity among user group members is also an important criterion (Loughhead *et al.*, 1994; Malhotra *et al.* n.d.). However, work by Chhetri and Pandey (1992) and Sarin (1993) indicates that this factor alone does not ensure success and indeed a homogenous group may be deeply divided in terms of individual dependence on public forest resources and thus interest in and incentive for protection of the forest. More important than the ethnic composition is the management and decision-making structure put in place. If all the interest groups (classified by gender, ethnic group, economic class, etc.) are fully represented and involved in decision-making and compensated if their livelihoods suffer as a result of forest management decisions, then, it is assumed, a fully functioning and effective local organisation will emerge. However, as the case studies indicate, this is often difficult to achieve and may require support and facilitation by outsiders.

Consensus about who should constitute the user group is probably one of the most critical factors in the development of a robust social organisation. If there is no agreement on membership of the group, there is little basis for developing management systems. Community forestry organisations in Nepal are formed only after thorough investigation of the users of a forest area; the next phase of implementation is not initiated until there is agreement on who should be members of the group. In the case of existing indigenous management systems, user identification is easier to achieve because there is already a recognised group of people accepted as users of a particular forest area (Nurse and Chhetri, 1992). The proportion of user households participating appears to be another important dimension to the functioning of the group. Studies have indicated that where there is a high degree of non-participation, the groups tend not to function well (Malhotra *et al.*, n.d.). It appears that non-participation usually hides conflict between different factions, and thus any activities the group undertakes may be sabotaged by the other factions.

In most of the JFM resolutions there is no provision for recognition of customary use rights, only for legally recognised rights. As Poffenberger and Singh (1992) emphasise, without the recognition of customary rights and their proper incorporation as part of a JFM agreement, it is unlikely that effective management of a forest will occur. Box 4.7 describes one example of what happens when customary right-holders are ignored in favour of those users defined by the legal framework.

Loughhead *et al.* (1994) found in a study of user groups in the Nepal-UK Community Forestry Project area that the original approach (criticised by some – see Soussan *et al.*, 1992) of taking the forest as the starting point has been eminently successful:

> ...the social institution of the group is in fact an essential prerequisite for the achievement of the programme's silvicultural objectives; the only way to identify user group members effectively is to take the forest as the starting point. This meant initially that the group formation process was slow – at the time of the Soussan report, only 36 had been formed – but by 1993 that figure had risen to 260, and group formation work is now in response to local requests prompted by interest in the programme and not necessarily as a response to resource scarcity.

In one case study described by Kafle (n.d.) he shows how important the initial process of user identification, negotiation and consensus-building can be in cases where the history of ownership of a particular forest area is contentious. In this one village development committee area Forest Department, local government and project staff tried to negotiate a consensus between extremely divergent and vociferous views. Although many commentators have suggested that such complex and contentious environments should be avoided, the experience suggests that it is important to try to resolve such conflicts if community forestry is going to have a secure future. In this case, the Forest Department took a strong role and threatened legal action against anyone causing damage to the forest while negotiations were still under way. Finally after one year of negotiation, agreement was reached and a user group formed.

Existing consensus arrangements

In cases where there are common management arrangements in existence, it has been found in both India and Nepal that forest management systems are more effective (Karki *et al.*, 1994). Although again the evidence on which this factor is based is still

Box 4.7 Users' and Forests' Boundaries: Who Is In and Who Is Out?

Where do you start the JFM process – in the village or in the forest? If you start with the forest, the end point may be very different from if you start from the village.

The diagram below explains this point in greater detail. As can be seen, the forest is dissected by a district boundary, and there are two villages which use the forest on a daily basis. Taking the forest as the starting point for the JFM process, the forest users are identified, both customary and legal right-holders: the users come from both villages. The JFM team then goes to each village and visits the users and discusses with them how they would like to manage the forest, recognising that there are two villages that have to be consulted. After a process of consultation in each village with all the users, the JFM team then brings the two villages together to decide how they are to jointly manage the forest resource. The villages decide it would be easier to split the forest in two and for each village to have its own clearly assigned area of forest. (The division of the forest was based on an existing agreement between the two villages where a ridge was recognised to be the boundary between each village's use area.) Together the two villages decide to formalise the boundaries for their forests, and agree that each village will only use its own area of forest. The villages each then set up their own Village Forest Committees and the management committees agree that they will meet on a regular basis to ensure that each village is continuing to respect the boundaries established.

Imagine a second scenario where the District Forest Officer decides that village 1 should be given management control over the forest because it lies within the revenue boundary of that village, and therefore legally villagers have the right to use the forest for certain products. The JFM team forms a VFC and management committee, and gives them the authority to exclude outsiders from using the forest. Village 2, in a different district and therefore outside the jurisdiction of the JFM team and without legal rights to use the forest, is not involved in the discussions and continue to exercise their customary rights to the forest as they were doing before initiation of the JFM process. However, now village 1 catches villagers from village 2 and tells them they are no longer allowed to use the forest since its management and use have been given solely to village 1. Village 2 users are angry and decide to cause as much destruction in the forest as possible, since they are no longer able to exercise their rights in the forest, and thus they no longer have a stake in the forest and its future production.

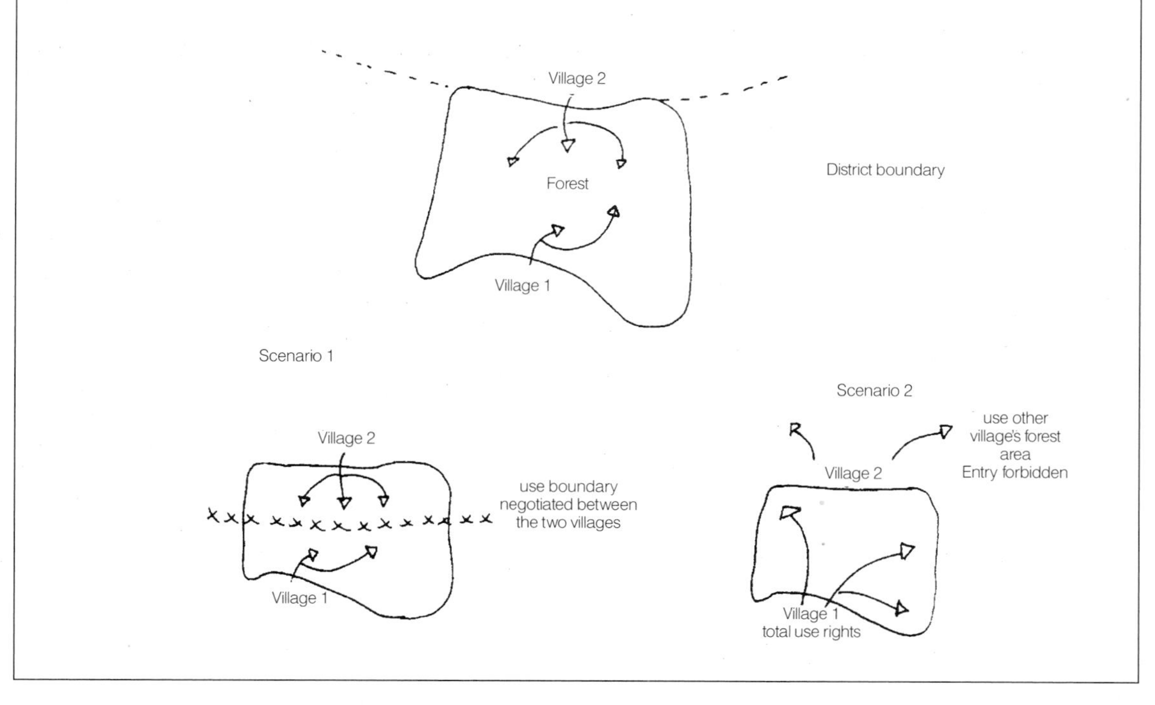

Strong local leadership is important for management of common pool resources, Nepal

quite limited, a detailed study of local organisations in Orissa came to the conclusion that without these strong village committees local action to protect forests would not have occurred so readily. The authors also suggest that in these areas the panchayats have relatively little power and thus have not disrupted the existing consensus systems (Kant *et al.*, 1991). Another example from Nepal shows the importance of village consensus systems where villagers are humbled in front of their peers when they have committed an offence against village rules. Box 4.8 describes this system. The importance of mutual relationships in controlling the problem of free riders cannot be overemphasised. This village assembly also monitors and audits the management of common pool resources in the village and determines the rules and procedures. In this case, since the assembly is mainly composed of wealthy men, it has been found that many of the rules do not favour the interests of women

Box 4.8 Peer Review: the Case of the 'Ghaunle Sabha'

The **ghaunle sabha** is a village assembly held every year. At the assembly each household is obliged to explain its activities over the year and to confess wrong-doings. It may take up to two weeks to hear every household's account.

The villager who confesses has to face a local jury consisting of village representatives. Each representative of the household has to swear in front of a copper bowl that their account is a true representation of events. The audience has the right to contest the account. After the report and confession, the jury declares the amount of the fine to be paid. Then the offender must pay the fine into the copper bowl. At the conclusion of the assembly, the jury collects all the money and sets the budget for the village council for the following year.

The social pressure exerted on villagers is great; the fear of losing face in front of the whole village acts as a strong deterrent to committing offences.

Source: Shimizu, 1994

Harvesting of chir pine plantation, Nepal

and poorer groups who depend on the forest for susbsistence (Shimizu, 1994).

The technology

From a review of the empirical material available it is noticeable that many of the most effective systems are based on formalised protection systems, as in Nepal with the **mana-pathi** allocation of grains to one appointed forest guard, and the rotational patrolling systems found in both India and Nepal known as the **tengapalli** system in Orissa and the **vara** system in Gujarat (Bahaguna, 1992; Kant *et al.*, 1991; Raju *et al.*, 1993; Tumbahampe, 1994, Box 4.9), in which group members take turns to keep watch during the day and night. Responsibility is often symbolised by a baton which is handed from one guard to the next (Femconsult, 1995c). In other studies in India payment using a similar system to the mana-pathi is also observed. In an investigation of forest protection organisations in Western Nepal a gradation of payments according to landholding was observed (Tumbahampe, 1994).

Box 4.9 The Staff and Red Turban

The system of **lauro** and **rato pheta** (staff and red turban) was instituted by the local leaders. Every day the villagers had to patrol the forests, each household taking a turn. The watchers used to carry a staff and wear a red turban, so that they were easily noticed. One watcher patrolled from one side of the forest and the other from the other side. They met in the middle and exchanged the staff and carried on walking across the forest. The staff was then left in the household who would be watching the forest the next day.

Source: Tumbahampe, 1994

Robust organisations have been identified where groups have invested considerable amounts of labour in active silvicultural systems (Bartlett *et al.*, 1992; Loughhead *et al.*, 1994). The robustness of groups that are solely protection-oriented should be questioned, as this often conceals institutional instability, with members having no confidence that their investment of labour will provide them with an assured return.

However, sustained investment in the resource will only occur when there is an assurance of a later flow of benefits. In many cases, local people have been unwilling to invest labour in the protection of forests because they are unsure whether in 10–15 years they will be allowed to harvest the benefits. Thus, the need to provide security of rights over produce is extremely important, together with the authority to exercise control over non-right-holders.

Congruence between use rules, local conditions and infringers

In most indigenous systems there are rules surrounding the use of forest products and their misappropriation. These are generally based on rationing access to products to particular periods of time and may use physical protection such as watchers, formal committees, or written rules that all members agree to abide by, such as the **gaja-patra** (a written list of rules) described by Malla (1992). The rules vary from group to group and are dependent on the type of product, demand and the ability of the group to impose sanctions. Box 4.10 describes an interesting system of graduated fines, with the ultimate sanction imposed on the group through the use of religious authority (see also Ingles, 1994). This is in contrast to the externally facilitated organisations under JFM in India, where duties and rules are prescribed in the government resolutions, leaving little flexibility for site-specific adaptations, and also in contradistinction to indigenous groups in India that have similar rules to those found in Nepal (Kant *et al.*, 1991; Raju *et al.*, 1993). An interesting point made by Kant *et al.* (1991) following their study of indigenous groups in Orissa is how organisations have diversified according to the particular local environment. Hence, in cases where there is little threat to the forest, use rules are relatively non-prescriptive. In other situations of extreme pressure, there are extremely elaborate use regulations and sanctions. This underlines the need to retain as much flexibility as possible within government guidelines and regulations to accommodate this diversity.

However, these rules can only continue to work as long as the users of the forest believe that there is some enforceable sanction. This is particularly the problem where groups have no legal basis and therefore can not effectively threaten users who transgress the group's rules nor do they have any power to deter outsiders from using the resource (SPWD, 1984; Kant *et al.*, 1991). The pressure from non-users or outsiders will also increase as the value of the protected resource increases.

Collective-choice arrangements

Studies in Nepal show that some of the most successful and robust user groups are those where operational plans have been modified in the light of experience (Bartlett *et al.*, 1992; Loughhead *et al.*, 1994). The converse to this has occurred in many user and JFM groups where members are unaware of their rights and responsibilities and have no vested interest in developing the operational plan.

Box 4.10 Regulation of Access

Karkiko Ban

In Karkiko Ban, Far West Nepal, access to the community-protected forest is regulated in a number of ways. Prayer flags (**neja**) from the Kedar Mandir are placed on the boundary lines of the forest and are renewed whenever illicit felling and abuse of the forest occurs. Religious beliefs – fear of punishment by the deity – prevent people from abusing the forest. Women are also prohibited from entering the forest during their monthly periods.

The Karki households confiscate the bamboo baskets (**doko**), ropes, and cutting implements from people who are found stealing products from the forest.

Seliko Ban

Felling green trees is restricted. If a family needs to build or repair a house, the committee issues permission to fell a green tree for timber. Any violators are liable for punishment in the form of cash fines. The villagers have maintained a common fund from this cash income, which is used to buy utensils for use during ceremonies, temple construction, or repair works. User households can also get loans from this fund at a reasonable rate of interest.

Someone violating the regulation for the first time generally pays a penalty of 5 rupees which goes up to Rs 50 and Rs 150 for the second and third violations. A fourth-time violator is tried in front of all the user members and they may fix any amount of fine deemed appropriate, depending on the seriousness of the violation.

Use of religious sanction

Following a series of infringements against the forest, the local people decided it was necessary to reassert control. A general meeting of the user members was called in a local temple to deal with cases of stealing. The people decided that each individual should take an oath. For this purpose, a bell from the temple was placed in the middle of the gathering. The household head or a member of the household had to swear (holding the bell in their hands), 'Neither I myself nor any member of my family have stolen forest products, including green leaf litter, from the forest. If we have done otherwise, may the deity punish my family.'

Source: Chhetri and Pandey 1992

In such cases the group tends not to function and the forests are neither protected nor managed (King *et al.*, 1990).

VSO volunteer experience from Nepal has shown that there are several key features to the development of effective user groups. These are described in Box 4.11.

Relationship between users and the state

These relationships vary from site to site, although there are some over-arching issues that emerge from JFM practice in India, where the presence of state functionaries within village forest organisations could be considered to be a means by which the Forest Department controls decision-making. There are extreme examples where village forest committee meetings are scheduled to ensure that forest staff can attend them, rather than being scheduled in accordance with local people's time commitments (Bahaguna, 1992).

The power retained by the state to disband user groups is common to both community forestry and JFM frameworks. It has been suggested by many that, although it is important for the state to be able to rescind agreements in case of violation, it is equally important for local groups to have some legal autonomy from the Forest Departments. In some States, organisations are registered separately, thus making it more difficult for the Forest Department to disband them (Poffenberger and Singh, 1992).

One particular case cited in North Bengal indicates the degree of

Box 4.11 User Group Formation

- Forest rangers attend user group meetings to give help and guidance, but not to design the programme.
- Forest user group assemblies are called to discuss rules for the management of the community forest and of conflicts.
- User group committee size should be between 8 and 13 people.
- All groups should be represented on the committee, including women and low caste groups. Women will often have different interests from men – their voices should be heard.
- Meetings held at 4 or 5 p.m. will allow women to attend after returning from their work.
- Women should be consulted about the rules governing collection of forest products.
- Women might not attend meetings. Men should be encouraged to discuss issues from the meetings with their wives and daughters.
- The rights of users who live far away or who occasionally use the forest should be discussed.
- Natural regeneration should be encouraged.
- Areas of forest that need thinning or felling should be described in the operational plan.
- Once a forest is handed over, the committee is responsible for issuing felling permits.

Source: Newsletter for VSO foresters

control retained by the Forest Department over local organisation membership.

> According to the (government) resolution, the Divisional Forest Officer concerned, in consultation with the 'Bon-O-Bhumi Sanskar Sthayee Samiti' (forest committee) of the Panchayat Samiti (panchayat committee) concerned shall select the beneficiaries who will constitute the FPC(s) ... Each FPC shall have an Executive Committee, comprising the Sabhapati or any member of the 'Bon-O-Bhumi Sanskar Sthayee Samiti' of the local Panchayat nominated by him, the Gram Pradhan or any member of the local Gram Panchayat(s) as nominated by him, and elected representatives of the beneficiaries (not more than 6) as members and the Beat Officer concerned as Member-Secretary. The constitution of the FPC including the executive committee must be approved by the Divisional Forest Officer concerned, on the recommendation of the 'Bon-O-Bhumi Sanskar Sthayee Samiti' of the concerned Panchayat Samiti'. (Roy *et al.*, n.d.:2)

As can be seen from this quotation, the Forest Department and formal administrative structures retain a large degree of control over the decision-making processes (see also Chatterji and Gulati, n.d.). However, one case cited by Bahaguna (1992) shows how some committees are prepared to fine forest staff when they transgress against the rules of the committee. In another case cited by Raju *et al.* (1993), due to the success of one village forest protection committee the Forest Department has withdrawn from the area. In Chapter 5, summary tables are provided of all the JFM agreements in India, and the majority indicate that the Forest Department retains a formal presence on each of the Forest Protection Committees. Under such conditions, it is difficult to envisage forest users having a genuine role within such an organisation.

The requirement in so many of the state resolutions for the Forest Department staff to be *ex-officio* members of user groups is also going to be difficult to fulfil if JFM is even moderately successful. In some cases groups plan to hold monthly meetings where it is required that the Beat Guard or another functionary should convene the meeting; the inevitable consequence is that meetings will not be held, and groups will become non-functional (Roy, 1992).

In many cases the composition of the Village Forest Committee is merely a formalisation of pre-existing relations between certain sections of village society and Forest Department officials. Often these relations have been negotiated over a long period and are mutually rewarding (Box 4.12). Although such organisations may not necessarily meet the criteria of equity and women's participation and empowerment, they do retain a large degree of stability since they do not challenge the *status quo*. In terms of sustainability they may survive into the future, even though they do not necessarily satisfy the requirements of forest development policy.

Beyond user group formation lies a whole area of support to

Box 4.12 Forest Guards and Villagers

In some villages, the SIFPGs have negotiated informal deals with their forest guards which recognise the villagers' contribution to forest regeneration. As the leaders of one SIFPG said, they recognise that the forest guard has to do his duty and cannot be asked to totally overlook unauthorised extraction. At the same time, the guard also has to accept that he could not function in the area without the villagers' co-operation. The 'deal' they have worked out is that, when any member needs timber for house construction, the group informs the forest guard when construction has begun. When the house is completed, the guard is invited to count the number of poles used and calculate the penalty payable. The guard, in turn, gives a 50% 'discount' by recording only 50% of the poles/timber actually used. This is in recognition of the villagers' contribution to forest protection.

Source: Sarin and SARTHI, 1994

ensure the continued growth and development of the group (Pokharel *et al.* 1993). There are three main aspects to this support: social – in terms of developing the internal institutional strength of the organisation; technical – the provision of appropriate and timely advice to ensure that the chosen silvicultural options meet the needs of the user groups; and physical – some small assistance with material inputs such as seed, stationery for maintaining records and partial assistance for forest watchers, although this latter support is usually considered to create dependency relations and to undermine the self-reliance of young organisations.

In order to develop the cohesion and bargaining power of local management organisations, there have been conscious attempts in Nepal and to a lesser extent in India to bring these groups together to form informal networks (Box 4.13). In Gujarat a group of Tree Growers Co-operative Societies, the Lok Van Kalyan Parishad, meets on the third Saturday of every month, and now has its own newsletter edited by a local NGO (VIKSAT). In Orissa there are a number of associations or fora of indigenous forest management groups. One large federation, co-ordinated by a coalition of grass-roots NGOs, includes 325 self-initiated community forest protection

Box 4.13 The Use of Study Tours to Encourage Experience Exchange

In Sunsari district in Nepal, the forest office encourages villagers to organise themselves into user groups. To help the users to understand the importance of forests and trees, and to guide them in their tree-planting activities, the forest office organises regular training and study tours. However, women rarely participate in these training programmes. In order to remedy this problem, the District Forest Officer organised a meeting with the women to discuss their training needs. As a result of this meeting, a study tour for women was organised. Each user group sent one woman on the tour. Each woman had the responsibility to pass on her experiences and the new knowledge gained during the study tour to the other user group members.

Source: VSO volunteer reports

groups. To provide nationwide communication, a JFM support network has been established in India, which links government, non-government and research organisations in an action, research and information-exchange framework.

Recent developments in Nepal have begun to develop the formal interface between user groups and Forest Department planning structures. Range-level planning and networking workshops at which user groups come together to share experiences and to plan now provide the information that forms the basis for the community forestry district plan. These planning fora first experimented with by donor projects have now been adopted by the government and have become institutionalised as range-post planning. This is discussed further in Chapter 7.

In yet other instances, with increasing institutional maturity some user groups have begun to function as local development organisations. For example, these groups have used funds obtained from selling forest produce to build schools. In other cases, they have decided to register as NGOs in order to gain access to development funds (Loughhead *et al.*, 1994). The recent formal national federation of user groups in Nepal is an important development in the strengthening of community forestry activities. This is discussed further in Chapter 7.

Conflict-resolution mechanisms

As has been described in earlier sections, conflicts arise over boundary demarcation, and recognition of customary as well as legal right-holders, between primary and secondary users, and between marginalised and non-marginalised groups. Recent work in Nepal indicates the primary areas of conflict to be:[4]

- identification of users
- sharing of benefits between households of different sizes and needs
- uneven participation of users in protection and management operations
- competition for leadership of the organisation
- conflict between user groups
- multiple rights in different forest areas
- conflict between forest user groups and the Forest Department
- conflict between national and local objectives

In all cases of successful local organisations, conflict-resolution mechanisms are used. In most of the above instances, conflicts can be resolved by negotiation, and it is rare that an outside agency is required to arbitrate. However, in some more serious disputes, Forest Department staff have played an important arbitration role (Box 4.14).

Box 4.14 Internal Conflict-Resolution Mechanisms

The most effective mechanism for conflict resolution of intra-village conflicts is frequent and regular meetings of the entire user group (not just the executive committee). Most autonomous forest protection groups in Bihar organise meetings where all members are obliged to attend. The penalty for non-attendance is loss of membership and the associated benefits. If a conflict cannot be resolved through open discussion, the leaders of the hamlets or other sub-groups are expected to negotiate. If these leaders cannot resolve the problem, the responsibility passes to a respected individual(s) whose decision is binding.

Source: Sarin, 1993

Government recognition of rights to organise

In both Nepal and India policy frameworks are supportive of participatory forestry approaches. Some anomalies have been noted where indigenous systems have not been recognised but have been supplanted by new organisations selected by external agents, often with disastrous consequences both for the functioning of the existing organisation and for the new endeavour. Indigenous systems have continued to be effective in cases where the government has recognised their presence and built on the existing structures (Kant *et al.*, 1991).

4.8 Decentralisation Policies: The Enabling Framework

The development of linkages between sectoral and political decentralisation is also an important part of ensuring sustained institutional change from bottom to top. In essence such linkages will help to provide a democratic forum through which the power of the line agencies may be challenged. The Forest Departments in both India and Nepal still retain a large amount of power and control over the village forest committees, indicating that the decentralisation process is only partial in its implementation. Currently, in India local forest organisations do not have any other institutional structure through which to question the actions of the Forest Department, or other line agencies; and in both India and Nepal the Forest Departments retain the right to dissolve the forest committees if they consider they have violated the agreement.

In India recent decentralisation activities within the forestry sector could be considered to have led to greater penetration of the state into the village, without the villagers acquiring an equal degree of power to question the actions of the state. As the state continues to reassert effective ownership over forest land through forest management organisations, the presence of forest officials in these groups is considered to be an essential controlling feature. In many situations, village forest committees established under joint forest management have effectively become an arm of the Forest

Department, rather than being developed as independent organisations that could challenge the authority of the department.

The example of the Panchayati Raj in India

As India moves towards a decentralised Panchayati Raj many people feel that forest protection and management should become the responsibility of the local panchayats as the lowest elected bodies representing the people. This has been supported through recent Government of India amendments to the Constitution (73rd Amendment) which give statutory authority to the village panchayats. In theory this implies that the panchayats could take responsibility for the management of forests within their boundaries. Similar authority has been given to district development committees in Nepal which have the power to form user groups and scrutinise and sanction sectoral plans (Wee and Jackson, 1993).

In India, however, several commentators have noted that, although the rhetoric points to lower-level decision-making, in actual fact 'with Panchayati Raj, the *power* of decision remains concentrated and centralised in the political and administrative hierarchies, though in form it seems dispersed through the various organs of local self-government' (Wade, 1988). Although this was a comment on the system 25 years ago, it is still considered to be the case in most States that the panchayat system does not decentralise control. Control is still mainly vested in the line agencies, and it is the relationships between the agents of the state and local people that determine where power is maintained:

> Officials are seen and see themselves as dispensers of favours. It is widely assumed that if an official wishes to do something for you he can, and the problem is how to make him want to. If you fail, it is because you do not have enough influence or have not paid enough money. (Wade, 1988).

The size of the panchayat will also militate against its use as the unit of management for forests; it is unlikely that organisations covering a diversity of social and ecological environments will be able to achieve consensus on resource management decisions (Poffenberger and Singh, 1992; Raju *et al.*, 1993; Bahaguna *et al.*, 1994). There are also additional questions about how representative of local people's needs the panchayats are (SPWD, 1992). Experience from Nepal is instructive in this respect. Early attempts to encourage local management of forests failed because they did not identify the actual users of the forests; rather it was assumed that handing management authority to the panchayat would lead to good representation of local people (King *et al.*, 1990; Hobley, 1990; Fisher, 1991).

However, as will be discussed later, unless there are close and coherent linkages between forest resource management organisations and the panchayat system it is doubtful whether these organisations can be sustained into the future.

The fate of the panchayat institutions is likely to play a crucial role in the future devolution of forest management responsibility, where the credibility of village-level forestry organisations will hinge on whether they are fully integrated into the political decision-making system. To date, there has been a large communication gap between the panchayat institutions and other village groups, with a lack of accountability and transparency about the allocation of panchayat funds to village development activities (Shankar, 1994). The panchayats have failed to represent the interests of the broad array of village groups, but equally the panchayats have not themselves been empowered. Control over decision-making has remained with the line agencies and politicians (Pal, 1994): paraphrasing Sanwal (1987) 'he who holds the budget holds the power'. Although budgetary control does not explain all facets of power ownership, it is considered to be of great importance within bureaucracies, where individuals are described as powerful because of their control over budget lines.

Many of the Indian States have shown great antipathy to the notions underpinning the panchayati raj system: 'at this level, politicians had little desire to create institutions that could provide alternative bases of political power to compete with or undermine their own' (Webster, 1990). Those States that did implement a programme of panchayati raj in the late 1960s (e.g. Gujarat and Maharashtra) soon terminated it as 'new centres of power and authority emerged' to challenge the government's authority (ibid.). The generally held view that the panchayat institutions merely reproduce and reinforce the existing power structures is contested by some who say that these institutions do provide an alternative structure through which groups can assert their democratic rights (Shiviah and Srivastava, 1990).

Some of the problems experienced with the introduction of the panchayat system and the reasons given for its failure (Box 4.15) apply equally well to the problems encountered with the newly formed forest protection committees under joint forest management.

Should this remarkable parallel in experience, more than 30 years apart, lead to disillusionment and dismissal of the possibility of creating more democratic institutions? The answer is probably no, although this more positive view should be tempered with a realistic assessment of the amount of time necessary to bring about change in local organisational processes, that necessitates the removal of top-down planning processes through a hierarchical forest administration to one where bottom-up planning integrates with local government structures and line agency delivery of services and support (see Chapter 7 for further discussion of this). This change also presents a fundamental challenge to socio-political structures at all levels.

In Nepal, new legislation enacted after the reinstatement of a democratic government has emphasised the important role of user groups, although it is not clear what the relationship should be between village development committees and forest user groups. This mirrors the state of uncertainty in India over connections between sectorally organised local resource management groups and the formal politico-administrative structure.

Box 4.15 Failure of the Gram Panchayat System
***Failure of JFM Organisations**

1. The ordinary villagers fail to distinguish between the **gram sabha** (village assembly) and the panchayat and are unaware of their rights and responsibilities as gram sabha members.

1* Forest users are often unaware of the rights and responsibilities conferred on them as part of the joint forest planning and management system. They do not consider they are able to challenge the decisions taken by members of the forest protection committees (elected representatives). The general assembly of forest users has no real decision-making power.

2. The nature of village politics is such that once a village leader is elected the villagers think that they have nothing to do after that and the leader will do everything. On the other hand, once an opposition leader is defeated, both he and his followers will cease to take interest in the gram sabha meetings.

2* Forest users often state that they have elected a chairman and committee to take decisions for them, although this contradicts the feelings expressed under point 1*. However, this is not unexpected, since the decision-making structures within a village are highly structured and exclusive. It could hardly be expected that in a short time the implementation of JFM would demolish these structures and construct new broad-based and open decision-making structures in its place.

3. When the gram sabha contains a number of villages, there is generally the lack of a common venue, which is easily accessible to the people of all the constituent villages.

3* This problem is found where the forest protection committee draws its membership from a number of villages. Problems of representation are greatly magnified, and for women, in particular, it becomes very difficult to travel to meeting locations remote from their household environment.

4. The timing of gram sabha meetings has much to do with popular participation.

4* Often meetings to discuss forest issues are held at times when the majority of forest users are busy with tasks elsewhere, thus reducing the opportunity to participate for an important section of the user group.

5. Very few people obtain information about the forthcoming gram sabha meetings. The usual method of communication is by the beating of drums by the village **chowkidars**, which is seldom done properly.

5* In the case of meetings called at the instigation of the Forest Department, usually a select few are notified (committee members). The meeting is often held at the convenience of FD staff, and in many instances meetings are cancelled without prior notice. This leads to frustration and irritation at the waste of time. Information and decisions from these meetings are rarely relayed to the rest of the forest user group.

Source: *Diwaker Committee Report on Gram Sabhas*, 1963 cited in Webster (1990)
*Additions drawn from field experience in India

4.9 Discussion

The material presented in this chapter represents many perspectives and approaches to developing local management arrangements for forests. However, since so much of what has been discussed so far under the umbrella of participation seems to fall far short of the ideal, what then are the indicators of a well functioning organisation? Again, these indicators depend on the perspective from which the organisation is being assessed. For example, an organisation dominated by a male elite could be considered to be successful from their perspective, although the groups without a voice may consider it to be a failure. A well-managed productive forest under the control of one autocratic leader may be considered to be eminently successful from a traditional forester's point of view; if, however, those whose livelihoods are dependent on access to the forest are being denied it and are suffering as a consequence, perhaps this is then a failure. Thus it is difficult to answer the 'winners and losers' question with any degree of confidence without stating from whose perspective the question is being asked.

Box 4.16 provides some indicators of institutional maturity which develop further the criteria presented earlier. These indicators will change as experience develops and the ability with it to measure maturity; they are comparative and not absolutes, and are presented here as an example of recent developments in this area.

In addition to these measures of group robustness, a further series of criteria can be applied to assess the form of participation (Box 4.17).

There are many positive changes in the forest sectors of both India and Nepal: new partnerships being formed, existing organisations being recognised and forests being regenerated. Nevertheless, the question of who wins and who loses has not been answered. To some these positive changes may be balanced by the increased penetration of the state into the village which began with social forestry and has continued with community forestry and joint forest management. Ironically, although decentralisation may force a greater transparency in certain relationships, it may also lead to increased linkage between the individual user and the state, a change which could be construed as increased centralisation of control and not decentralisation.

Box 4.16 Indicators of Institutional Maturity

- Number of groups formed (although this provides no measure of quality)
- Number of members and drop-out rate
- Community agreement on membership
- Frequency of, and attendance at, meetings
- Attendance of women at meetings
- Percentage of women on committee or other decision-making body
- Number of groups forming cluster links with others
- Attendance of group members at leadership and skills training workshops
- Clarity in, and understanding of, roles, responsibilities and relationships
- Members' labour and material contributions to group activities
- Users take responsibility for forest protection without externally funded forest watchers
- Democratic changes in leadership over time (elections) – although some groups prefer to select rather than to elect
- Consensual production of microplan and implementation workplans
- Negotiated access to other forest areas for products not supplied in own forest
- Evidence that work plans have been adhered to and the specified outputs achieved
- Consensual revision of microplans and work plans in the light of experience without external support
- Evidence of conflict resolution without recourse to external arbitration
- Effective application of skills to maintain group assets (i.e. pruning, singling, planting, weeding, etc.)
- Mutual support between group members for other non-project activities
- Examples of collective bargaining with local elites
- Ability to access external agencies for support and services as required

Source: ODA, 1994; KOSEVEG, 1994; Chhetri *et al.*, 1993; Karki *et al.*, 1994; Raju *et al.*, 1993

Box 4.17 Criteria for Assessing Quality of Participation

- *transparency*: whether all stages of the activities are publicly visible, including decision-making processes
- *access to information*: whether there has been adequate and timely access to relevant policy and project information
- *accountability*: whether the agencies involved in management and implementation are procedurally and periodically answerable to people
- *meaningful choice*: whether people can participate in a voluntary manner without being compelled, constrained or otherwise left with no other choice
- *comprehensiveness*: whether people have been consulted from the outset in defining the nature of the problem or opportunity prior to any project being decided upon, as contrasted with consultation during subsequent stages of the project cycle
- *non-alienation*: whether people have participated in such a way as not to feel distanced and alienated from development activities, the implementation process and the eventual outcomes.

Source: Adnan *et al.*, 1992 quoted in Bass *et al.*, 1995

Notes

1. Where common property regimes are described as 'institutional arrangements for the cooperative (shared, joint, collective) use, management and sometimes ownership of natural resources' ... common pool resources refer to the physical qualities of a natural resource and to the social institutions human beings have attached to them. 'Common property resource is avoided because it conflates property (a social institution) with resources (a part of the natural world)' (McKean, 1995).
2. Runge (1986), Ciriacy-Wantrup and Bishop (1975), Ostrom (1990 and 1994). These are some of the key texts. There is otherwise a large and expanding empirical literature. The mainly anecdotal evidence underpinning these assertions is now being tested through a large longitudinal research programme co-ordinated across the world by Elinor Ostrom at the University of Indiana.
3. The conditions under which collective management systems operate in Nepal and India are also well documented (see for example, Arnold and Campbell, 1986; Dani *et al.*, 1987; Hobley, 1990; Sarin, 1993).
4. For detailed case-study descriptions of these different conflict situations see a recent edition of the *Banko Janakari* (1995).

5 Who Benefits?

Everybody listens to the rich but no-one listens to the poor even if they shout. If some project comes to the village, the rich go to the front and take their share of the money: the poor get nothing. (Sano Sunaar, quoted in Hobley, 1990)

The foreigners told the villagers that because women are the forest users they must also be members of the forest committee . . . The problem cannot be solved by outsiders imposing such ideas on men. If men wish to dominate women then that is what will happen. (Kancha Chetri (a woman), quoted in Hobley, 1990)

5.1 Introduction

Critically, in the eyes of the donor community participatory forestry has been funded with the objective of ensuring that poor people dependent on forests have greater control over the sources of their livelihoods. This chapter questions this objective in the light of experience, and asks how far participatory forestry can or does address the needs of marginalised groups within society. The evidence presented looks at the formation of individual relationships that each user experiences, and tries to provide a context for understanding how individual users are affected by development interventions.

The material is organised around the question of 'who benefits'. It looks at the relationships between the different actors involved in participatory forestry and the balance of power between them. The list below indicates the levels of relationships, some of which are analysed in this chapter through case-study material, while others – those between donor and implementing agency, and within the implementing agency and between government and non-governmental partners – are considered in Chapter 7:

- Donor and implementing agency
- Within the implementing agency
- Between government and non-governmental partners
- Between implementing agency and village
- Between villages
- Between close and distant users
- Between forest-dependent users and those who are partially dependent
- Within the village: between powerful and marginalised
- Between legal right-holders and customary right-holders
- Within households
- Between resource-rich and resource-poor areas

Chapter 4 looked at the diversity of interest groups and individuals involved in participatory forest management and provided examples of some of the emerging partnerships. This series of studies looks more critically at the power relations between individuals.

> The decision-making apparatus may be dominated by outsiders to the village community who may form part of the elite. They may be government officials or even well-intentioned activists, but they are still outsiders. Even if the decision-makers are part of the village community, what does one mean by the term community?[1] Are the decision-makers the elite of the village, the rich, the powerful, exploiters in one way or another? Even if the decision-makers represent the very poor, the disadvantaged, the oppressed, they may perhaps be representing only men in a male-dominated village society.
> (Fernandes and Tandon (1981))

5.2 Incentives for Forest Management

If institutional change is predicated on the development of effective local organisations, what will ensure that these organisations continue to function into the future? At the centre of this question lie incentives that will encourage structural as well as behavioural change. Although some of the characteristics of robust organisations have been examined in Chapter 4, what has not yet been examined is the incentives for individuals to act collectively for the common and not individual interest.

This question of incentives is extremely important in areas where local people already have good access to forest products under previous arrangements, such as **nistar** rights and forest settlements (as in Himachal Pradesh). In some cases where *de facto* use of forest resources has enabled local people to retain 100% of forest products, the joint forest management sharing arrangements are perceived to be unattractive. For many marginalised groups, however, the 'hassle' factor experienced by villagers involved in illicit activities is high, and for these users legalised use of forest resources is attractive. This is particularly apparent in areas of relatively high forest cover. In the best situation villagers are allowed to retain 50% of the proceeds and in the worst scenario only 25%. Even factoring in additional payments that must be made between villagers and functionaries, villagers probably had greater real access to forests and their products prior to the introduction of JFM (Kolavalli, 1995).

In situations where *de jure* use rights allow villagers good legal access to forest products, JFM offers even less. In these cases, villagers are being asked to exercise their rights with responsibility, and in some circumstances to curtail their use. This exercise entails attendance at meetings, donating household labour for protection functions and in some cases allowing non-right-holders access to the resource. For those who may currently be disallowed access to forests, participatory forestry may provide an opportunity to gain use rights. In some cases, it may be used by those who already retain social control within a group to increase their power further (Hobley, 1990).

In addition to the incentives provided or withdrawn through the formalisation of access rights, there are other questions about costs and benefits that need to be addressed (drawn from Uphoff, 1992):

- Time: do benefits/costs accrue rapidly? If not what is the impact?[2]

- Space: benefits/costs accrue locally rather than remotely. What are the implications of this for those who do not share in these benefits (particularly those traditional users who are excluded by geographical distance)?

- Tangibility: benefits/costs are evident and are not hard to identify

- Distribution: benefits accrue to the same people who bear the costs of management rather than to different people. This is an important issue with respect to women's participation and control over use rights.

- Those who were benefiting before more formal systems were put in place are not disenfranchised. This relates to marginalised groups within the village, and those distant from the forest.

In the following sections, these questions will be considered in more detail, looking in particular at how marginalised groups are involved in participatory forestry and the impact on them.

5.3 The Non-Formal Institutions

What is the village? The role of the individual in collective action

The new philosophy talks about the devolution of power and control to 'local people'. However, it rarely disaggregates this term to its constituent parts. Who are the 'local people', what is a 'village' and who are the 'users'? Without a correct identification and clear understanding of the client group, it is unlikely that local forest management organisations will be sustained over the long-term. Problems are arising in both India and Nepal where it is assumed that committees are representative of all the users, and thus it is not necessary to have more generally representative bodies such as a general assembly. In these cases, the interests of different users are often ignored, with only the interests of the most dominant being heard (see also Thomas-Slayter, 1994).

> The principal function of a local institution in JFM is to provide an institutional structure which can articulate and represent the interests of *all* user sub-groups of a forest area in the partnership agreement with the FD.
>
> (Sarin, 1993)

Recent experience points to both the potentiality of local organisations and also their frailty. This frailty can be induced by either endogenous or exogenous factors. In particular, organisations break down because of the non-inclusion of traditional users of a resource, as was discussed in Chapter 4.

This is less of an issue where there are pre-existing resource management organisations, but this may then lead to the question of how equitable these organisations are. This question is of importance to donors in the forestry sector who may place a high premium on the participation of marginalised groups in forest management organisations, and indeed to policy-makers who make explicit reference to fulfilling the needs of poor people and women.

In Nepal a far more careful critique has been carried out of this issue, and there is a large amount of operational experience to indicate ways in which the further marginalisation and disempowerment of poor people can be avoided (Box 5.1). In India, however, many of these questions are only now being asked, and

Box 5.1 How the Needs of the Poor Are Ignored

When the process of user group formation has been rushed or short cuts attempted it has often been the poor who have suffered. This has happened, for example, where initial user identification has been poorly carried out, where individual household visits were replaced by group meetings and when people who had been missed on the first visit were not followed up subsequently.

The poor are not accustomed to being asked to take decisions. Often there was a 'psychological subservience' that hampered their speaking out. Where there was no prior attempt to build up their confidence, poor people were excluded by their silence.

Time given to general assemblies was often too short to reach consensus. If consensus had not been reached beforehand, the proposals of the group who could shout the loudest or gather a majority would be passed. Likewise, assemblies that passed rules on a majority rather than a consensus basis worked to the detriment of the 'silent' poor.

Often the institutional arrangements of the user group worked against poor people. For example, if they were not represented on the user group committee they might not hear about a forthcoming assembly. Committee members elected because they were 'personalities' rather than as representatives of a particular group of households often did not represent poor people.

Some management plan prescriptions aimed at improving forest management acted to the detriment of poor people. For example, long rotation lengths, priority for timber production, closing the forest for long periods of the year, and banning grazing and charcoal production all have a direct negative impact on poorer groups whose livelihoods are dependent on full and uninterrupted access to forests.

Source: Julian Gayfer, pers.comm.

practices in many instances have not led to the improvement of poor people's livelihoods.

Aid projects are at the interface between Western development ideology – with its insistence on the empowerment of the individual – and other ideologies that may insist on the subjugation of the individual. For example, as Wade (1988) describes from his research in South India, 'territorially-defined groups like villages are not a focus for [Indian villagers'] identity and needs. Indeed, the strength of attachment to non-territorial groups like the sub-caste is said to obstruct emotional attachment to the village.' The village, in an Indian context, is better seen as a group of individuals tied in a series of horizontal and vertical patron-client relationships that extend up into the state hierarchy (Wade, 1988; Pathak, 1994). Any intervention in these relationships therefore should be viewed in this context.

This important insight should be tested against the rhetoric that states that local forest management should be organised at the 'village' level. The question then becomes whether the village is an appropriate institution through which to implement such programmes, and one with which villagers themselves identify, or whether there is some other grouping that better represents the ways in which local people view the forest resource and its management. We shall return to this question later on when we consider who are the users of the resource, and who should benefit from local forest management.

Are we asking too much, therefore, when we state that these village organisations should be representative of all groups? Perhaps this is denying the cultural strictures that regulate village and external interaction through the mediation of a few members of certain caste and gender groups. Indeed, as villages and households become increasingly integrated into external economies, the process of disaggregation and differentiation also increases (Cernea, 1992).

How equitable are indigenous organisations?

Experience and empirical evidence point to the effectiveness of indigenous organisations as the basic unit from which to build improved forest management organisations. However, as has already been alluded to in Chapter 4, there are equity and representation questions concerning the internal processes of these groups.

At one extreme lie those groups formed by local leaders following exposure to the user group concept through workshops or hearing about it at district meetings. It appears in these cases that the process of formation has led to many problems – users are poorly identified, there is little or no consultation in the development of the operational plan, no consensus over which area of forest is to be managed. These types of organisations seem to have some of the worst attributes of both indigenous and externally

Box 5.2 Equity, Conflict, Representation and Indigenous Organisations

- villagers do not want to include users from outside their political boundary
- neighbours are excluded because of political factions and/or ancestral control of the forest
- conflicts between local groups in terms of their claim over the forest boundary – particularly if the forest extends over a large area
- distant users are included where there is a kinship relation
- most of the important decisions are made by committee members; women and poor people are ignored
- timber extraction costs are charged, and timber auctioned – problematic for poor people
- the groups are more protection – than utilisation-oriented
- the younger generation is less interested in protecting the forest
- Department financial support, particularly for plantation establishment and watchers, has created dependence and heightened expectations

Source: J.C. Baral, 1993; Tumbahampe, 1994

facilitated organisations, with few of the benefits of either (Drona, 1994).

This raises the issue of the role of external facilitators (such as aid workers but also any individual external to the village context) and the degree to which they can interfere with existing social structures to mould the organisations to meet the objectives of an outside funding organisation. Box 5.2 indicates the outcome of an analysis of groups that have been formed by local leaders with little or no external facilitation.

Some have contended that a minimalist approach should be adopted where a policy of 'if it looks good, leave it alone' is the most effective (Fisher *et al.*, 1989). Others, accepting that indigenously derived organisations often have many positive attributes but rarely fulfil criteria such as empowering women or allowing poorer people a voice in decision-taking suggest that, in such cases, external facilitation is appropriate (Box 5.3 and Sarin, 1993).

Box 5.3 Structure of User Committees

The local committees are often not representative of all groups within the social hierarchy and can in some instances merely help to reinforce the status of already established local elites. The tendency of extension staff is to initiate mass meetings to decide upon the election of a users' committee. Unfortunately at such meetings only the powerful and dominant individuals tend to speak. Therefore smaller interest groups or women's groups tend to be marginalised. The use of female extension staff helps to alleviate these types of problems, since they are more able to converse with local women and draw them into discussions.

Source: VSO Volunteer Reports

Blacksmiths whose livelihoods are dependent on regular access to forests for firewood, Himachal Pradesh, India.

Who has user rights?

In order to address the issues of institutional sustainability and the impact of decentralisation, it is essential to understand who exactly are the users of forest resources, and what are the implications of changing forest management structures for those who are excluded from access to the forests. This returns us to an analysis of whether it is possible to ensure that all users can be given rights to forest resources through the new property rights institutions being promoted by donors and activists alike.

Between forest-dependent and partially dependent users

This question of who is a user is extremely complex. In a sense if the question is approached from a livelihood perspective (instead of from the 'recognition of traditional rights perspective'), the answer is quite different. In this instance, an individual whose sole source of livelihood is derived from forest products would be identified as a primary user, instead of the definition provided in Chapter 4 based on proximity to and usage of the forest (see Box 5.4). Priority would be placed on groups such as headloaders (whose only source of household income is obtained from sale of firewood) over and above other groups in a village who may partially secure their household needs from the forest, but may also obtain some tree products from private land, and thus, in general, their livelihoods are only partially derived from the forest.

Between customary and legal right-holders

Throw into this complex equation the rights of indigenous communities who have been displaced from forest areas by incomers, and the whole situation becomes one of multi-tiered negotiation, both spatially and temporally. Indeed, the initial problem is to identify who has a legitimate claim to the benefit of the resource, and who

Box 5.4 Is Proximity the Only Criterion? An Example from Haryana, India

Residence in geographic proximity to a forest area is a major factor weighing in favour of practical involvement and commitment to sustainable management. However, residence alone may often be inadequate as an eligibility criterion for user group or local institution membership. With the expansion of the Haryana programme, in many cases the residents of adjoining and, at times, fairly distant villages started protesting against joint management agreements which had been exclusively negotiated with only one of the numerous traditional user groups. They claimed equal, if not greater, traditional rights in the same forest area as those closer to it.

Case diagnosis of such conflict situations has revealed a diverse variety of relationships between user groups and forests. The only one of these formally recognised by the Forest Department is the rights or 'concessions' granted to villagers under various forest settlements. In many areas, however, although no rights are specified in the forest settlement, people still claim rights in well-defined areas according to an earlier revenue settlement. While not recognised by the FD, these rights are accepted and honoured among village groups and effectively determine which groups have access to a particular forest tract to meet their subsistence needs. By overlooking such tacit arrangements, forest departments may inadvertently deprive one group of its traditional access to a local resource by legitimising the exclusive access of another group. In addition to increasing inequity between different users, this may also sow the seeds of inter-group conflict where none existed before.

Source: Sarin, 1993

Box 5.5 Who Are the Users?

Prior to the mass settlement of large areas of the Tarai, it was primarily inhabited by the Tharu people and other related forest-dwelling groups, including Majhis, Rajbansis, Satars and Darais (Bista, 1987). The Satars are a landless semi-nomadic group who were dependent on the forest for their livelihood, both as a source of food through hunting and gathering, and as a source of shelter. With the rapid disappearance of the forests in these districts, these small groups have come under intense pressure and have become increasingly marginalised and exploited by incoming hill groups.

According to community forestry policy in Nepal, forests should be handed over to those who have traditional rights to use them. In this case, the traditional users were too far distant to benefit from the policy, and indeed lost the right to collect products when adjacent forest communities took over protection of the forests and banned other users' access.

Source: Hobley (1992a)

determines what is or is not legitimate. As the example presented in Box 5.5 indicates, legal right alone should not determine who has managerial control; customary rights in this case have been practised for generations and should not be ignored just because they are not legally recognised. In this example, the traditional right-holders have effectively been disenfranchised through the process of community forestry.

Just as in Nepal, some users have also become disenfranchised in India. The proponents of joint forest management sometimes appear blind to the social, ecological and political diversity of the nation, and apply the model irrespective of the location (although the consequences of the application are highly diverse). In the example from India, in Box 5.6, traditional users who are no longer dependent on the forests are not prepared to allow new settlers whose livelihoods are dependent on access to forest products to enter into JFM arrangements.

Many local organisations are being established in a fashion that takes little cognisance of local imperatives, and will not lead, either in the short or long term, to sustainably managed forests. There are

Box 5.6 Livelihoods Versus Rights

Jholuwal is a large and fairly prosperous Jat village in the Pinjore range in Haryana. Although the Jats claim traditional rights in the adjoining forest compartment, they depend on it only marginally to supplement fodder available from their private landholdings. In contrast, a more recently settled community of Banjaras in Moinawali, living closer to the forest, has primary dependence on **bhabbar** grass for earning their livelihood through rope-making. During exploratory negotiations initiated by Haryana's JFM support team, the Jats refused to accept that the **bhabbar** grass lease should be sold to the Banjaras while they purchased the fodder grass lease for the forest compartment. They wanted the **bhabbar** grass lease to be sold to them on the strength of their traditional rights. As a consensus could not be reached on Moinawali's right to participate in the joint management agreement, no agreement could be finalised.

Source: Sarin, 1993

even some examples where there are two forest protection committees set up in the same village by two different parts of the Forest Department, one under social forestry, the other under new JFM arrangements. They may have the same membership but the rules and benefits governing association are entirely different (Chaffey *et al.*, 1992).

Within village relationships between powerful and marginalised: the role of women as users

At the heart of decentralisation and participatory forest management policies lies the endeavour to provide formalised property rights to those groups who may previously have had only informal or customary access to the forests, access which was practised at the whim of forestry officials. In addition, many development projects following these new forms of forestry aim to bring marginalised groups into the development process. Accordingly, projects emphasise the involvement of women in decision-making about resource allocation, and assert that it is those whose livelihoods depend on the use of forest resources who should have authority over management decisions (Molnar and Schreiber, 1989; Siddiqi, 1989; DN, 1990; see also Tinker, 1994 for a discussion of the role of projects in promoting women's participation). Since in most cases the major collectors and users of forest products for both domestic consumption and sale are women, any intervention in the forest management system is going to have the greatest impact on women.[3] For example, it is estimated that 2–3 million people are engaged in the headloading of fuelwood in India, the majority of them being women (Venkateswaran, 1994).

Property rights are, however, highly gender-specific and should be considered in several dimensions including control and use (Joekes *et al.*, 1994). Thus, institutions may be established that allow local people to manage resources, but the key target groups may still remain partially excluded from process. This is particularly the case for many women, where their use rights may be secured through this process but they still have no control over the management of their rights (Loughhead *et al.*, 1994). Control remains vested with the male members of the group. The question of how women are incorporated into the development process through forestry programmes is still incompletely addressed, although it has been high on the project agenda for most of the last 10–15 years. Webster's (1990) study of general participation in panchayat organisations also reveals a similar series of fundamental problems with promoting women's participation, to those encountered in forestry:

> [Women] are rarely present at the public meetings ... The idea of participation by women in any kind of meeting is rarely considered by men and laughed at by women ... Attendance at a meeting would also imply that women had a role in decision-meeting which most men consider not to be the case ... The women from the more affluent households never attend these

> meetings, this is both a gender and caste phenomenon. (I)t is not merely a question of being elected but of being able to assert a presence within the meetings as well. Caste, gender and the prevailing norms of social behaviour with respect to elders and the educated remain as obstacles here.

Women's views
Men don't care about the forest. They go to the teashop, drink tea and talk, that's all they do. Women cook food and feed them. Men don't have to cook so they don't have any concern about the forest . . . It is women who need the forest, they need firewood to cook. Women can't advise men, even if they do the men don't listen to women . . . Men preach to women about not cutting trees, but what can women do? They cannot cook food without firewood and they cannot collect firewood from other places.
Men's views
Women cook and eat rice; they do not go to committee meetings.
(Group discussion with women in Kabhre Palanchok and Tamang men in Sindhu Palchok, quoted in Hobley, 1990)

This was also the case found by the author in Nepal, where research carried out on women's participation in the community forestry decision-making process indicated that projects aiming to increase women's involvement often actually increase social tensions between men and women, and may lead to a reduced role for women:

> The foreigners told the villagers that because women are the forest users they must also be members of the forest committee. According to the foreigners it should be compulsory for women to attend the meetings. The men agreed to this and women were allowed to become committee members. However, women were informed of a meeting only when a male committee member chanced to meet them. Even if women attend meetings they cannot voice their opinions: they cannot speak against the opinions of their seniors. When the men have finished speaking that is the end of the meeting ... Men do not tell women that they cannot speak at the meetings, but the men do not want to be opposed by women. (Kancha Chhetri quoted in Hobley, 1990)

In many cases, women are not allowed to attend meetings where senior male members of their household are present (Chatterjee, n.d.). Even if they do attend they are reluctant to speak out because they are afraid of making mistakes; they think the men will laugh at them (Drona, 1994). 'The important thing is that men should realise the importance of women's views regarding forest manage-

Separate men and women's groups to discuss their objectives for forest management, Nepal

ment. The problem cannot be solved by outsiders imposing such ideas on men. If the men wish to dominate women then that is what will happen.' (Sama Chetri quoted in Hobley, 1991).

However, in the face of men's opposition to women's attendance, it is difficult for women to speak out at meetings. One woman stated that if women do speak at meetings men will say that 'the hen has started crowing'. The following quote from one villager indicates a generally held view about the role of women in public meetings and reinforces the latter point – that to expect a radical change in social structures as a result of participatory forestry is far from the reality of village life.

> There are women on the committee; there is one woman from every household. Whether women are called to the meetings or not depends on the amount of work at home. They are called to the meeting if their participation in the meeting is urgent ... Calling all the women to the meeting just hampers the progress with the agenda because discussion is not substantial ... it is true that women are the real users of the forest but our women have not yet participated in the meetings. They don't know much, they can't give solid opinions. Let me tell you one thing, I am a man, I attend the meeting. If I am prepared to make the female members of my family act according to what I say, why should they attend the meeting? (Bahadur Chhetri quoted in Hobley, 1990)

The reasons for the silence of women are thus complex and deeply embedded in their social relationships, and it is naive to assume that participatory forestry alone will lead to fundamental changes in these gendered relations, or indeed any other participatory development approach (see Jackson, 1993 for further discussion of these issues). It also leads the practitioner to question the validity of approaches that insist on the involvement of women, irrespective of their interest in the issues or their ability to articulate their views in a male arena (Mosse, 1993; Shrestha and Drona, 1994). Box 5.7 presents an interesting example of women using their relationships with men outside their households to maintain their rights of usage to a forest that is now under protection by a village committee. This reveals the conflict between different groups' objectives and the satisfaction of livelihood needs.

The choice of methods, timing of meetings, location and formality are all aspects that must be considered in order to enhance and facilitate the involvement of women, within a context of understanding the social relations that govern a woman's interaction at different stages in her life.

Within households: the role of women

To understand more clearly the relationships that govern men's and women's interactions in public fora, it is helpful to consider some examples of how these relationships are constructed.[4]

There is a hierarchy within households dependent on the age

Box 5.7 Women as Forest Offenders

In some situations male members of forest protection committees are unable to take any action against women who continue to remove products from a protected forest for fear of the consequences from their husbands.

Dukhu Tudu, an old man of Dahi village in Midnapore district of West Bengal, once came upon a few women who were involved in indiscriminate cutting of trees in the nearby forest. Dukhu tried to prevent them from damaging the forest but the women refused to listen to him. He tried to stop them by snatching away their axes but, in turn, it was he who was blamed for misconduct. The women complained to their husbands about Dukhu, accusing him of physical attack and molestation. The husbands then came in a group seeking revenge, with the result that Dukhu Tudu suffered serious head injuries.

Source: Chatterjee and Roy, 1994

and gender of the individual. Women have many roles and relationships that will change over time, from woman as daughter to woman as wife to the head of the household. These relationships are more highly developed within joint households where there is greater conflict between married women and unmarried daughters of the house competing for the same resources. These tensions are often observable at village meetings where both are present but where only the mother-in-law will speak, for example.

A woman as wife to the head of a joint household has considerable control over the allocation of work amongst junior wives who are married to her sons; she will delegate household tasks and will make many of the internal household decisions. A woman as daughter of the house holds a favoured position often with considerable freedom of action; she is not obliged to carry out dirty or defiling tasks, which are allotted to the most junior wife who will be expected to wash the plates after the meal and remove any night soil.

A woman as wife also has a particular relationship to her husband which conditions the way in which she responds to other men, and also how she may behave in public whilst with her husband. The deference a wife must show to her husband permeates all aspects of her life and is clearly expressed in the public arena of public meetings and village decision-making.

Rules regarding access to forests may be constructed around the prohibition of women to forests during their menses. Women are considered to be a source of ritual pollution during their menstrual period. A woman's relationships with other men, and in some circumstances other women, are heavily circumscribed: a woman is not allowed to prepare or touch food and water, she may not enter the kitchen areas and must eat separately from the rest of the household (see Box 5.8).

Thus every time a woman attends a public meeting her response to that meeting and to the other participants is conditioned by these relationships. However, if women are not consulted and their voices not heard, the form of forest intervention will often reflect only

Box 5.8 Impurity and the Removal of Women from Domestic Life

During Satibama's menses she could not touch or be touched by other people. She remained outside the house until nightfall and then came indoors but had to sit apart from us, she was not allowed to sit on the same mat or anywhere near the cooking or warming fires. This latter fire is the second fire which is lit in the evening to keep people warm; everyone sits around this fire except for the daughter-in-law who had just washed the dishes and was sitting by the cooking stove. The warming fire is open with three large flat stones forming a circle; it is fed with small twigs and branches and is outside the ritually pure area of the cooking fire. Satibama was not permitted to go upstairs to her bedroom but went to sleep with the animals at her aunt's house. The next morning she returned to the house after her pre-sunrise wash, but was not allowed indoors, she remained outside and kept her distance from us. She could not eat with us nor be served by other people, she ate after we had finished and washed her own dishes.

Segregation is strictly observed and women refer to this period of the month as **na chune** (i.e. not touching).

Source: Hobley, 1990

male priorities which may not coincide with those of women (Box 5.9).

The degree to which women are involved in participatory forest management thus varies according to a complexity of social and cultural factors. The evidence to date shows that in both India and Nepal women's participation is relatively limited (for example, in one forest division in West Bengal only 2% of the total forest protection committee members are women (Roy *et al.*, n.d.). In Madhya Pradesh, women's lack of involvement is directly recognised and separate women's forest committees have been estab-

Box 5.9 Agroforestry Training for Farmers

Agroforestry forms a complementary part of the community forestry programme and is usually included in most user group activities. An interesting example from the Tarai indicates the real differences between men and women's priorities and land-use choices, again emphasising the need to be not just site-specific but also to take full account of each user's particular needs.

Participants for the agroforestry training course were selected only from active user groups and one-third of the participants were women. As part of the three-day training, the last afternoon was spent in a group exercise. The participants were divided into three groups, one all-women group and the other two groups all men. Their task was to design an agroforestry system for their own farm. There were interesting differences between the groups. The women primarily considered their family needs. They included a large diversity of plants: vegetables, fruit-trees, spices, trees for fuelwood, fodder and timber, grasses for fodder, and medicinal plants for household use. The men regarded it as a commercial exercise. In their systems trees for timber production were the primary concern, with spices and medicinal plants as additional sources of income.

Source: VSO Volunteer Reports

lished. Although this may be the least challenging way of tackling the problem of low levels of female participation, it does not address the central question of how to increase the number of women actively involved in village forest committees (Narain, 1994). Other States have now amended their JFM resolutions to state explicitly that two members from each household (male and female) should be members of the general body, and in some circumstances have also increased the number of women on village forest committees. However, as rural social patterns change with increasing numbers of women-only or women-headed households, government policies will also need to respond accordingly, placing more overt focus on means to bring these groups into the decision-making arena (Box 5.10 and Lingam, 1994; Sarin, 1993).

In Haryana, the government resolution has been amended to allow full adult membership as opposed to one male one female member from each household. This decision was taken to prevent intra-household problems where younger women could be excluded from participation because their interests would be considered to be represented by the senior woman of the household. This does, however, raise other problems, such as whether there are too many members for effective participation in decision-making (Sarin, 1993).

As many families are extended or joint families with three or four generations represented, selection of one man and one woman could pose problems. In the case of the eligible female member, options could include the mother-in-law, elder or younger daughter-in-law, unmarried daughter, an unofficial second wife, a widowed or abandoned daughter, etc. Each would respond differently in a public arena and represents a different perspective according to their status and security of tenure within the household.
Who should decide which man and which woman should become members and on what basis?

(adapted from Sarin, 1993)

Forest-dependent groups

For many poorer households the forest provides the major source of livelihood requirements; any change in access to these resources therefore has profound consequences for them. Since these groups also tend to be the most marginalised and least likely to be included in decision-making about forest management options, they are the most vulnerable to radical changes such as closing the forest to allow regeneration. It is estimated that in India over 30 million people are entirely dependent on the forest for their livelihoods (this includes artisan groups such as basket-makers and potters). Similarly common property grazing areas (often the target of participatory forestry programmes) are also heavily depended upon by poor people to supply grazing and firewood needs (69% and 91%

Box 5.10 A Women's User Group

As a result of the work of one woman VSO volunteer, the rangers with whom she worked decided it was important to involve women in user groups and aimed to achieve 50% female membership of the groups. However, the success of this approach has been patchy, particularly since the women tend not to trust male forest workers and have little self-confidence in the presence of men in official positions. There have been, however, some interesting successes. For example in one all-women user group, where no natural leaders emerged, it was the only group where political factionalism also did not occur. It is not clear, however, whether these two observations are directly related.

Source: adapted from VSO Volunteer Reports

Basket-makers, Bihar, India

respectively) (Agarwal, 1992).

Increasingly, forest management groups are assigning a monetary value to forest products that were previously free (Drona, 1994). In the words of one poor man, this decision taken by the few but affecting the livelihoods of many will have profound consequences for poorer households:

> This has caused a distinct line between poor and rich. The poor don't have money, what can they buy firewood with? The rich

'If the villagers own the forest they may protect it or they may destroy it. If the watcher is from the village, he may allow only the rich people to use the forest and not the poor people . . . If there is a government watcher, he may treat everyone equally.'
(Sano Sunaar quoted in Hobley, 1990)

have money so they will buy firewood. The poor always suffer in all aspects from every side. If a rich man's small hut is damaged, he can apply for a tree. He can then cut down a big tree and build a house. If a poor man needs to cut down a tree to build a house, they advise him to ask other villagers for a tree. If a powerful person needs a tree for some small job he can easily apply for it. Ram Bahadur is an example. The wall of his house was slightly cracked. He brought people from the forest office and they cut down some trees. The poor cannot give chickens to government personnel, and hence they do not get any trees. (Sano Sunaar, quoted in Hobley, 1990)

Other groups disadvantaged by the devolution of management authority to settled villagers are the nomadic graziers. They have the double problem of being seasonal and non-resident users so that they are rarely present when forest protection groups are being set up, and their homes are distant from where they exercise their rights. As described in Chapter 4, conflicts with settled agriculturalists are an increasingly common feature of nomadic life, although there is a long recorded history of conflict that has been the subject of much debate, particularly in States such as Himachal Pradesh, where some of the major nomadic grazing routes occur (Tucker, 1986b). Many adminstrators have looked for solutions to these conflicts; most have centred around negotiating individual access to forest lands or reduction in flock sizes (Rizvi, 1994). In order to reduce conflict now, it is going to be necessary for the settled users to negotiate with the graziers some form of access system which could be based on a system of permits for a certain number of livestock to be determined on the basis of fodder availability (as is currently organised by the Forest Department). It is also possible that, since it would be difficult for the graziers to negotiate separately with a large number of forest protection groups, this negotiation role could be taken up by the Forest Department on their behalf. As yet there have been no workable solutions to this difficult problem, but it is suggested that without some form of negotiation joint forest management in these areas will be characterised by conflict between settled and nomadic livelihood systems.

In the following case-study provided from the Tarai the effects of forest protection on other dependent groups is considered.

Natural forest is highly valuable and can meet more than just the demands of local people. Why should one of the country's most important resources be distributed to a limited number of rural people? It can produce highly valuable timber which is in great demand on the world market. If managed well, it can become an important source of income for the nation.
(VSO volunteer reports)

5.4 High Value Forest: a Local or a National Resource?

As the implications of participatory forest management are more clearly articulated, the question for whom these forests are to be managed, now has to be answered. In the case of high value forests, the allocation of immediate and long-term values to local people has major implications for the nation as a whole, where effectively a direct transfer of benefits is to be made to a group of people whose geographical good fortune places them in a position

The Tarai landscape: a village in the interior of a forest area, Nepal

to reap the rewards of a model which is not necessarily appropriate to these areas.

Although there are many positive experiences from the hills of Nepal, in the plains (the Tarai) the situation is quite different. Here the application of the community forestry programme has met with many problems, as described in Box 5.11. This also reflects some of the experiences in India in areas of high value resources with good access to markets (Chambers *et al.*, 1989).

Regional appropriateness of participatory forestry: the experience from the Nepal Tarai

Although state services may appear to have had little impact on rural Nepal, the state has long played a major role in structuring rural society. Despite difficult communications, the quasi-feudal system consolidated under the Ranas supported a hierarchical and exploitative rural order, many features of which survived under the system of panchayat government established in the 1960s and persist today, as is evidenced by the labour relations prevalent in the western districts of the Tarai. This is a major reason why the state, in its relatively recently established role as a provider of services, has so far failed to deliver significant benefits to much of the rural population. Frequently, such benefits have been captured by rural elites, leaving little to filter down to others. Like other development initiatives sponsored by the state, community forestry in the Tarai runs the risk of falling prey to local elites.

In the Tarai many areas are under relatively unstable farming systems because of the uncertainty over tenure, the illegality of land occupation, and tenancy relations which do not lend themselves to investment in long-term land development (Box 5.12). For most people, land-holdings are small and divided into many plots.

Box 5.11 Issues in Community Forestry in the Tarai

- strong market forces and high commercial value of forest products
- acute land-use conflicts
- low dependence on forests for fuelwood and fodder
- young settlements and acute ethnic diversity
- wide gap in trust between the local people and forestry staff
- authoritarian attitude of field staff
- frequent transfer of DFOs and other key staff
- poor communication between the centre and the field
- inadequate extension activities
- difficulty in identifying the real users
- discontinuity of programme
- tendency of user groups to demand forest areas for harvesting and not for long-term sustainable management

Source: Shrestha and Budhathoki, 1993

Fragmentation and declining household size have been an increasing problem over time, with a growing proportion of households becoming landless or land-poor (Soussan *et al.*, 1991).

In areas where there has been migration from the Hills, the farming systems tend to mirror the hills system, with greater use of fodder trees and fuelwood. In the southern areas of the Tarai, there is no tradition of growing or using fodder trees, and the traditional preferred fuel is dung sticks.

Farming households are by no means restricted to agriculture to provide their livelihoods, and many households supplement their incomes by seeking alternative employment. Much of this employment centres around illegal trade in firewood and timber, with landless households most dependent on this trade to supplement their income (Timsina and Poudel, 1992). In some cases the collection of non-wood forest products such as leaves, grasses and medicinal herbs also provides another valuable source of income. Several detailed studies (Subedi *et al.*, 1991; Muller-Boker, 1991; Timsina and Poudel, 1992) have shown the importance, both direct and indirect, of different forest products to particular local econo-

Box 5.12 Tenure Insecurity and Implications for the Community Forestry Programme

- landlords are unwilling to make productive fixed investments (such as trees, irrigation)
- tenants are rotated frequently, removing their incentives to make sustained improvement to the land, such as planting trees
- fixed assets such as trees remain the property of the landlord and not the tenant, although the tenant may have some usufruct rights over the products
- landlords prefer to use their own household labour to farm small plots less intensively than is optimal, rather than using tenants who may claim land ownership
- landlords prefer to use Indian labourers who will not make claims on the land, thus increasing the social disaggregation prevalent in the Tarai

Source: adapted from World Bank, 1991b

Dung sticks, the preferred fuel in the southern Tarai, Nepal

mies. A study carried out with a Tharu village in Chitawan revealed that they gathered 62 different plant species, the tubers, leaves or fruits of which were consumed; 36 plant species were used in house construction, and 63 as medicinal plants (Muller-Boker, 1991).

As well as the local users of forest products, increasing urbanisation and tourism are placing additional and unsustainable demand on forest resources. A number of industries also use large quantities of fuelwood, for example brick kilns, distilleries, and tobacco barns;

Tharu village, note the heavy use of wood in construction of housing and carts, Tarai, Nepal

smaller enterprises, such as restaurants, bakeries and tile makers, also consume significant quantities of fuelwood (Soussan *et al.*, 1991). Much of the small industry demand is fulfilled through headloaders with little or no other source of income and livelihood support. Thus it can be seen that any change in access control will have severe and direct consequences on some of the most vulnerable groups in society.

The landless rural people have no alternative source of forest products other than that available from community land or national forests. Their livelihoods are often entirely derived from forest lands (Subedi *et al.*, 1991). Although in many respects these groups are the ones who should receive priority attention, they are generally the ones excluded from participation in community forestry. They are the most 'invisible' group mainly because they are rarely present in the village during the day because they are busy grazing cattle, collecting fuelwood, share-cropping or engaged in some other activity by which to support their households. Even if they do attend meetings, it is often difficult for them to speak out, particularly if it is in contradiction to the views of their landlords:

> Most of the Tharus in Dang-Deukhuri have been very greatly exploited by zemindars, landlords and revenue agents. They are virtually slaves in the hands of the zemindars, sold and bought at will. (Bista, 1987)

This practice of bonded labour still occurs in some western districts of the Tarai, and restricts to a large extent the development of the full agricultural and forestry potential of the area. The Tharu labour force has little incentive to invest in long-term management of resources, including trees on their landlords' farms, or protection of state forest resources. There is a risk that these particularly vulnerable communities could be made more vulnerable if their access to

Bauhinia leaf plate collectors' temporary camp, Tarai, Nepal

Box 5.13 Bauhinia Leaf Plate Collectors

This case illustrates the problems involved in identifying all users of forest products for one particular area of forest. Unlike in the Hills of Nepal, where users are relatively easily identified and tend to be associated with one geographical area, in the Tarai users may come from different districts.

We came across a temporary camp of 35 men, women and children who had come to this area of natural forest for a period of 15 days. They lived in Mahottari district, and came to collect leaves to make into leaf plates during the agricultural slack season. Each year the same group of people come to this area and pay the Forest Department Rs 400 for a permit. They say there are several groups of leaf collectors who go to other forests in other districts. Previously they used to be able to use a nearby forest but it has now become a wildlife reserve, and consequently access is now banned, even though, as they say, their collection practices are non-destructive.

The group expects to collect 15 bundles of leaves over 15 days. Each bundle contains 20 units of 1,000 plates each . . . They say that they can sell 100 plates for Rs 5. The gross proceeds from their leaf collection is Rs15,000. However, costs of transport, permit and other incidental payments should be taken into account. The plates are finished by the women when they return home and they sell them in the nearest large town.

They say it is now more difficult to find places to collect leaves than it was five years ago. This is partly because forests are disappearing but also because they have become protected as wildlife reserves or national parks. They also said that they used to collect fuelwood and leaves from Sagarnath forest before the natural vegetation was replaced by a eucalyptus plantation. This used to be their preferred area for leaf collection.

Summary:

- highlights the importance of non-wood forest products as an income supplement for poor people
- the importance of recognising seasonal user rights
- the difficulty of identifying seasonal users, particularly when they live at a distance from the forest
- the consequences of changed management practice on forest users, as was shown by the Sagarnath example.

Source: Hobley, 1992

forest land is not protected. In areas where the landlords control labour, it is unlikely that user groups composed of this labour force will be able to retain sole control over community forestry resources, and it is conceivable that their access to community forestry might be less than if these same forests remained under state control.

Since these people's livelihoods often depend entirely on the forests, their knowledge of the use and harvesting systems for different products is often detailed and extensive. However, it is their very dependence that also leads to their often exploitative relations with the forests, and makes them one of the most difficult groups to work with effectively through participatory forest management programmes. Once forests are brought under local protection this usually entails restricting access to those people whose livelihoods are most dependent on retaining secure access to forest products (as illustrated in Box 5.13).

In these socially and ecologically complex and difficult areas, community forestry has been attempted with limited success. The World Bank supported a community forestry project from 1983 to 1990, with subsequent additional periods of funding. Practical implementation of the programme was supported through a series of expatriate volunteers derived from different international volunteer programmes, including VSO and Peace Corps volunteers. They highlighted some of the difficulties there are in trying to implement community programmes where the local society is highly fragmen-

Box 5.14 The Tarai Community Forestry Project

Objectives:

- to increase supplies of fuelwood, poles, timber and fodder by establishing community forests, farm and other private plantations, strip plantations, woodlots, large block plantations, and by improvement of natural forests
- to introduce, popularise and distribute improved cooking stoves as a fuelwood conservation measure
- to expand training facilities to meet the additional need for technical and professional staff; and
- to strengthen forestry research, field operations, and planning activities

Project problems

- frequent staff transfers
- late budget release which affected timely implementation of programmes
- inappropriate stoves (low rate of installation and use – in the best case stoves were used as flower pots)
- too much control at centre with little delegation of authority
- staff shortages. One assistant ranger could not cover one **ilaka** (local administrative unit) of 50–100,000 people effectively
- training provided to field staff was outdated
- the community facilitators formed user groups too rapidly in order to obtain cash incentives

Source: Hobley, 1992; VSO Volunteer Reports

ted, differentiated and politically manipulated. In essence, one of the major problems with the implementation of community forestry in the Tarai has been the inappropriateness of an approach derived from experience in a totally different social and ecological context – the Hills (Pardo, 1993). Box 5.14 provides a summary overview of the objectives of the programme and some of the subsequent problems.

In the name of community forestry and enhancing access to tree products, **taungya** systems were established on degraded Forest Department land. However, experience has shown that the many negative social aspects far outweigh any biological gains through increasing tree cover. This was a programme specifically targeted at the landless groups. These households were allocated 0.5 and 1.5 ha of land and were paid to plant trees. In between the trees they were allowed to cultivate the land for a maximum of four years, at the end of which time the trees had grown to such an extent that agricultural cropping became virtually impossible. At this stage the landless households were moved on to another area of Forest Department land and the process was repeated. Box 5.15 indicates some of the problems encountered with the taungya systems.

In conjunction with the introduction of taungya as a means to protect the natural forest areas, community plantations were also established, again with mixed success. In general, the approach used has experienced many of the problems faced in India under the social forestry programmes, where plantations were established by the Forest Department with little local involvement. In the Tarai, plantations established under the community forestry programme were supposed to be handed over to local communities, except in

Box 5.15 Taungya and its Problems

- there were no formal procedures for the selection of genuinely landless participants
- distribution of land was not on the basis of need but of relationships between the 'landless' settlers and project staff
- plantation land was difficult to protect since it was distant from the settlement
- insufficient land was allocated to ensure adequate levels of agricultural production, and much of the land was not cultivable; this forced household members to look for work outside the area
- there was no access to agricultural extension advice
- it provided a cheap form of protection for the Forest Department but limited benefits to the landless settlers
- short-term agreement (4 years) led to insecurity and no investment in local infrastructure, such as water supplies
- the types of agricultural crops allowed were restricted to those that would not interfere with tree growth; the settlers were not allowed to prune the trees to prevent shading of their crops
- the expectation of permanent settlement of the land was not fulfilled; the participants had hoped that this would give them the opportunity to become 'legal'

Source: Shrestha and Pandey, 1989

some situations where the plantations, established with user involvement, were on land designated as national forest, in which case users would have no right of access or usufruct. The plantations were used as a visual marker to indicate Forest Department ownership of land and a means by which to deter further encroachment. In most cases there was little local ownership developed and therefore little interest in the protection of these tree assets (Box 5.16).

In one case, where the plantation was to be handed over to the local people, the DFO complained of the difficulties caused where two factions had formed – one from each political party. In addition, the District Chairman had also interfered because he wanted the composition of the users' committee changed to reflect his own political allegiance.

As Boxes 5.17 and 5.18 indicate, there have been some successful experiences, and the reasons for success provide valuable lessons for future practice.

Exploring the continuum: the need for different institutional arrangements in high value forest areas

As experience has shown, the models developed in a low value, socially stable hills environment are totally inappropriate to most parts of the Tarai (Box 5.19). These high value forests with diverse and competing user demands require a different form of management, where much of the control remains with the Forest Department and where leasehold forestry potentially has a role (Sharma, 1993). This raises an interesting issue about the conditions for different types of property regimes; in some cases where it is

Box 5.16 Failed Community Forestry Plantations

Rautahat District

This village had a 50 ha area of plantation established two years previously. However, there was minimal survival of seedlings since most had been grazed. There was little commitment amongst the villagers or interest in the fate of the plantation. Since all the labour had been paid for by the Forest Department, and the watcher was also paid by the FD, the village had no stake in the plantation.

Lessons learnt:

- no commitment from the users
- negative impact of DOF paid watchers
- insufficient time was given to community forestry by the rangers, with limited follow-up support to the group
- user group did not understand what community forestry meant
- poor representation of users – a small all-male committee
- political factionalism, one political party blamed the other for destruction of the plantation
- user group was formed at a time of political unrest
- only a small number of users were consulted

Box 5.17 Conditions for Success

Case study 1

Chitwan District: river training

Members of one village development committee have come together to protect and plant an area (105 ha) of degraded riverside, which was previously used for grazing. Their motivation was several-fold: to protect their agricultural land by planting trees to prevent the river changing course, and to provide a future supply of timber and firewood to alleviate their current firewood problems.

The villagers had heard via the radio and by reading posters that there was a community forestry programme in their district and approached the District Forest Office for assistance; they also visited neighbouring areas to look at other community forestry activities.

The user group committee, composed of 15 people, was elected at a full village meeting with one representative from each household, in the presence of the Forest Department. On average the committee meets 3–4 times each month to discuss protection issues. The users also meet about once a month to discuss how to use the money collected from the sale of grass, how to protect the trees, and how to market the sand and rocks from the river.

Protection of the plantation from grazing cattle was a major problem; initially, it was difficult to protect the plantation area from the cattle of their own village. It was only after the first year of protection when significant quantities of thatching grass had regrown that villagers began to see the benefits. This was reinforced after the village had organised an auction of the grass to a contractor. The contractor paid for the rights to harvest the grass – Rs 24,000 – and this money has been deposited in the users' group account; he then employs the villagers to cut grass for him. In addition, villagers are allowed to cut grass on the area not bought by the contractor, for their own use, at a price of Rs 40. They have a ticket system, one ticket allows one household to cut as much grass as is possible within a certain period. If they cannot cut sufficient in the time allotted they are allowed to buy another ticket.

The users estimate that they should have about Rs 200,000 from the sale of the grass. They plan to invest half this money in better protection of their plantation and to use some of their grass money to pay for forest watchers (since there are no longer any project funds available as the project has ended).

The villagers have also, as a result of reaping the benefits of protection, decided to change their livestock grazing practices. Now most cattle are stall-fed with grass cut from the plantation area.

Lessons learned

- importance of local identification of the need for a plantation
- role of extension materials and local demonstration to raise awareness
- broad-based consultation and consensus
- importance of an early intermediate product – in this case grass
- the ability of users to pay for watchers with revenue from early intermediate products
- adequate fodder supplies allowed a change in livestock management practice
- early benefits led to an enhanced commitment to protect forest land

Source: Hobley, 1992

Box 5.18 Conditions for Success

Case study 2

On the southern slopes of the Churia hills rising out of the plains of the Tarai, a village has been protecting a 90ha area of natural forest and plantation, including 15 ha of bamboo. It is the only forest in the area with some remaining *Acacia catechu*. However, because the forest has not yet been handed over to the villagers, they have very little authority to prevent outsiders from using it.

About 10–12 years previously this Tharu village had become very concerned about the erosion problems in its area, with gullies opening up and deepening. As a response to this the villagers formed a committee, to try to protect the Churia by stopping other villagers cutting firewood.

Their attempts gained real impetus after they came to hear about the Community Forestry Project. The villagers asked for their protection committee to be broadened to form a users' group. At first the village was divided into several groups; some wanted to protect and manage an area of forest, whereas others were not interested. Several members of the user group were taken on a study tour which helped to show them what other groups were doing. Study tours facilitated by VSO volunteers in other parts of the Project area had an important role in raising awareness, creating networks between user groups and demonstrating new approaches (see Chapter 7, Box 7.11).

Prior to forming the user group, the villagers first formed sub-groups of interested villagers; each sub-group then elected a representative to the committee. Women had their own sub-group and also elected 5 members to the committee. Women users spoken to were actively involved in the committee, and were well aware of the decisions taken.

The committee meets regularly to discuss forest-related problems, and all the users are called together at least once a year to decide what to do with the forest. They are planning to sell firewood to the users for a nominal fee. Already they sell bamboo and **banmara** (*Eupatorium odoratum*.) However, this group is frustrated in its efforts to manage the forest, as the DFO has not been able to hand it over to them.

They have now successfully protected the forest for over a decade. Livestock grazing is prohibited, and they impose fines on those who transgress this rule. They say that their protection has stopped firewood cutters from coming to the area and indeed the condition of the forest is such that it did not appear as if there was much outside disturbance.

Several important factors are contributing to the success of this initiative:

- awareness of erosion problems led the villagers to form a committee with the objective of stopping outsiders cutting firewood and increasing forest degradation and erosion
- importance of using local field staff who speak the local language to ensure more effective extension
- effective local protection has stopped firewood cutters from coming to the area
- the villagers started by protecting a small patch of forest and then slowly extended the boundaries as they became more confident of their ability to protect the area
- there were clearly defined and agreed forest boundaries using natural features such as gullies
- high-level of awareness in the village about the importance of forests, created through study tours and because several members of the village are in the forest service
- no obvious factions within the village – it is a homogeneous Tharu village
- forest is easy to protect because it is close to the village
- committed Forest Department staff, very supportive of the work of the villagers
- Users have become their own extension agents: women from this group tell villagers in their paternal homes about their community forestry activities

Source: Hobley, 1992

not possible to divide the resource into units of manageable size, it may not be possible for local groups to protect the resource from outsiders (McKean, 1995). This reinforces the notion that it may be the case that common property regimes may not be appropriate to all situations.

Some of the experiences in India with joint forest management are instructive for the Tarai, where sliding scales of benefit-sharing have been introduced according to the value of the forest. The following ideas from a group of Indian foresters were suggested to

Natural forest protected by the adjoining village in the Churia Hills, Nepal

Box 5.19 Hills Versus Tarai: Some Realities

The Hills

1. Coinciding with the ecological heterogeneity of the hill environment, small patches of forest are scattered irregularly all over the hills.
2. Villages are also scattered and generally nucleated and associated with certain patches of forest for many generations. Due to the inaccessibility and limited economic value of these forest patches, the interest of outsiders is also limited and fairly manageable.
3. Villagers experience the direct relationship between degeneration of the forest cover and its impact on their livelihoods.
4. Local protection of forests leads to obvious benefits to the protector.

The Tarai

1. The northern part of the Tarai, with its gravelly and dry Bhabar soils unsuitable for agriculture, is covered with the remnants of a large valuable sal forest. The southern part of the Tarai is relatively devoid of forests, although there are more private woodlands, and trees on farms.
2. These sal forests are intensively used by people from all over the Tarai, from the Hills and also from India. Everyone is interested in this highly accessible and valuable forest. Forests are under pressure from uncontrolled livestock grazing, squatting and cutting for fuelwood, fodder, animal bedding and timber. This is exacerbated by businessmen engaged in smuggling fuelwood and timber to Kathmandu and India, and by politicians who encourage encroachment and squatting with promises of legal land registration in return for votes.
3. The forest cover has decreased dramatically. The causal understanding seen in the Hills still appears to be missing in the Tarai, where transport is readily available to move forest products from areas of relative abundance to scarcity areas.
4. The age-long associations of particular settlements to one forest area seen in the Hills is absent in the Tarai together with the responsibility and accountability for the trees in that forest. Thus, if you don't cut the tree, your neighbour will.

Source: VSO Volunteer Reports

help develop a form of community forestry more appropriate to these socially and politically complex areas:

- sharing the output of Tarai community forests between user groups and the government on a 50:50 basis or 75:25 basis (in contrast to the hills where 100% goes to the users)

- sharing the financial input provided by the government between individuals and the group as a whole

- self-financing versus joint financing, i.e. the amount of investment by the different parties becomes the basis for calculating returns to each party (similar to buying shares in the forest)

- cash income could also be distributed to individual users, as in West Bengal

- a certain percentage of the products could be kept for urban needs

- small-scale industry should be developed in association with management of these forests (Campbell and Denholm, 1993)

Again this reinforces the need to retain flexibility within policy and legislation in order to ensure that site-specific responses can be made to diverse environments. Chambers *et al.* (1989) provide a useful comparison between villages in the hills and in the plains, indicating some of the reasons why collective action may not be so easy to catalyse in plains villages (Table 5.1).

Table 5.1 Comparison between upland and plains villages and factors affecting potential for collective action

Characteristics	Upland/tribal villages	Villages in the plains
Topography	• undulating	• flat
Population	• 50–100 families	• 500–100 families
Type of wastelands	• mainly owned by FD	• mainly revenue and private
Use	• one village	• several villages
Dependence on wastelands	• very high	• alternatives available
Authority of village elders	• still intact	• only within caste group
Market and state penetration	• weak	• fairly strong
Caste homogeneity	• one caste dominant	• multi-caste
Minimum requirement for regeneration	• protection only	• funds required because of degree of degradation
Users within a village	• all families	• mainly the poor

Source: Chambers *et al.* 1989

5.5 Who controls the Partnerships

In the analysis of joint forest management resolutions presented below (Tables 5.2 and 5.3) it is instructive to look at who controls the partnerships. As can be seen in all cases, the Forest Department retains firm control over the formation and dissolution of the village organisations (Kolavalli, 1995). In the main, representation on these organisations is heavily prescribed by the Forest Departments, with the majority of positions given to non-elected members, often outsiders to the village. Women's representation is again prescribed but, as has been seen from the case studies, prescription rarely leads to active participation.

5.6 The Benefits

So far we have considered the costs to some of the groups involved in participatory forestry. However, there are obviously major benefits for groups involved in the management of forests, particularly in areas where forests regenerate relatively easily and there is a ready market for forest products. In such cases, local groups have already received large amounts of money from the sale of both intermediate and final products, increasing the group interest and incentive to maintain the forests as a means to enhance their local developmental environment (Box 5.20).

Just as the harvest period can be one of great excitement, it can equally be a period of great disappointment when villagers find that their years of protection are not repaid since the value of the final harvest is significantly less than they had expected. This was the unfortunate experience in Arabari, where after the State Forest Corporation had harvested the final crop and extracted its manage-

Distributing the harvest of a chir pine plantation, Nepal

Box 5.20 The Harvest

In Soliya in Gujarat, a gram vikas mandal (GVM) created under an AKRSP project, the forest they had been protecting was finally harvested. In 1992, some 56,000 bamboo poles and 1,700 eucalytpus poles were harvested under the technical guidance of the Forest Department. A 25% share of the bamboos was given to the GVM plus the whole proceeds from the eucalytpus harvest since it was grown on revenue and not Forest Department land. The net revenue from the sale of the bamboo and eucalyptus amounted to Rs186,000. However, there was some dissatisfaction expressed by the villagers who said that the Forest Department should not have received such a large proportion of the benefits (75%) for doing nothing.

In addition to these benefits there had been an increased flow of non-wood forest products such as seeds, leaves, flowers and fruits for local consumption. In some cases, villagers had sold the surplus to an agent in the village, bringing in an additional source of income to the households.

Source: Balooni and Singh, 1993

ment costs (53%), very little money (12% of the gross) was left to divide amongst the users who had protected the forest for over 14 years (Poffenberger and Singh, 1992). Even before this experience the people of Arabari had had to wait 15 years for a formal resolution on benefit-sharing, meanwhile expending time and labour in the protection of a resource from which they had no guarantee of gaining benefit (Chambers *et al.*, 1989).

These success and failure stories reveal several important factors:

- since the time span for harvesting forest products is relatively long-term, it is difficult to predict whether the markets will be available for the products or not

- it is important to have a flow and diversity of intermediate as well as final products to ensure the maintenance of interest in forest management and to reduce the risk associated with managing a resource for one product

- the principle of net versus gross proceeds should be clearly explained to groups, and where possible the groups should be responsible for contracting the harvesting services of the state so that the costs associated with harvesting are transparent to them

- the success of participatory forestry depends on the type of resource available for protection. Many areas available for JFM do not have the ecological vigour necessary to provide an immediate flow of benefits of sufficiently high value or utility

- generation of forest income is affected by the ratio of forest area protected to the number of participating households, the biological productivity and the amount of labour required to protect and manage a given area of forest (Poffenberger and Singh, 1992)

Table 5.2 Analysis of joint forest management resolutions

State & date of JFM resolution	Legal status & name of organisation	Eligibility for membership of General Body	Composition of MC/EC & Women's membership
Andhra Pradesh 1992	Registered by FD, no autonomous status, **Vana Samrakshana Samiti**	one male one female adult per household	Forester mem.sec, FG, president of GP, 6–10 elected reps • at least 3 women
Bihar 1990	Not stated bound to follow PCCF's orders Village Forest Management & Protection Society	one rep. per household	Min. 15, max. 18 members composition specified (1 mukhiya, ex-mukhiya, sarpanch, teacher). Forest watcher mem.sec • Min. 3, max. 5 women
Gujarat 1991	Gram panchayat or registered cooperative soc. **Van Samiti**	no prescription	Working committee of GP rep, 2 women, 1 NGO rep other interested members • Min. 2 women
Haryana 1990	Autonomous registered society. Hill Resource Management Society	all male and female adults	Decided by HRMS general body through annual elections • Min. 2, provision for all women committee
Himachal Pradesh 1993	Registered by FD no autonomous legal status Village Forest Development Committee	one male one female per household and president of MM, YM, & GP member to be nominated by DFO	One rep per 10–20 adults; FG ex-officio mem.sec.; Member of GP; Rep Antoyodya h'hold; Member of MM • 50% women
Jammu & Kashmir 1992	None constituted by FD Village (Rehabilitation of Degraded Forests) Committee Village Plantation (Protection & Management) Committee	one adult male/female per household	Forester or FG mem.sec. min. 2 women and 2 SC/ST/BC • Min. 2 women
Karnataka 1993	Association registered under Karnataka Societies Act by DFO Village Forest Committee	One representative per household Up to 10 ex-officio members specified by FD	Forester ex-off mem.sec. 10 elected reps including 2 SC/ST, 2 women, 1 landless labourer, plus 4 nom. members • 2 women

Table 5.2 (continued)

State & date of JFM resolution	Legal status & name of organisation	Eligibility for membership of General Body	Composition of MC/EC & Women's membership
Madhya Pradesh 1991	None, constituted by FD Village Forest Protection Committee	one representative per household	One rep. per 10 households, all GP & Antyodaya committee members. Ex-officio: RFO mem.sec, village kotwar, teacher, chief of village • Not specified
Maharashtra 1992	Can be registered FLCs or FPC constituted by FD	one representative per household	Forester mem.sec., sarpanch, gram sevak, 6 elected members (2 women, 2 ST/SC/BC) • 2 women
Orissa 1993	None, constituted by FD, known as Vana Samrakshana Samiti	one male/female per household (to include all who exercise rights)	10–15 members; sarpanch, ward member (s), 6–8 elected/selected reps (3 to be women), forester mem.sec., forest guard, NGO rep. (selected by DFO) • Min. 3 women
Punjab 1993	None, constituted by DFO. Forest Protection Committee	none specified, no general body	Sarpanch, members of village panchayat, FG, VLW, three selected villagers (SC, one woman, one ex-serviceman or other ex-govt. employee) • One woman
Rajasthan 1991	Autonomous cooperative soc. eg Tree Growers Coop., Forest Labour Coop, Village Forest Conservation & Devt. Soc.	All residents	No provision • No provision
Tripura 1991	No legal status. Const by DFO. Forest Protection & Regeneration Committee	one person per household selected by DFO (only households with one wage earner eligible)	FG mem.sec 5 elected reps • No provision
West Bengal 1989	No auton. status, reg'd by FD	Joint membership of husband & wife DFO to select members in consultation with PS	FG mem.sec; Reps of PS & GP, max. 6 elected reps. • No provision

Sources: Sarin, 1993; SPWD, 1993; Femconsult, 1995 a,b,c

Table 5.3 Power and benefits in joint forest management

State	Power to dissolve local organisation & final deciding authority	Benefit-sharing NWFPs & Timber
Andhra Pradesh	DFO can disband committee only after consultation with District Social Forestry Committee • Final decision : conservator of forests	All NWFPs free of cost except those leased and auctioned. Rights for collection vested exclusively with Samithi members Timber and poles, 25% of products for own consumption; 75% retained by FD and one-third of income earned credited to Samithi
Bihar	Local infringements by members can be resolved by other members of Society • Final arbiter not described	Dry branches, grass, leaves, other produce available at market price One-third share of timber income for village development fund
Gujarat	DFO can cancel agreement • Final decision: conservator of forests	Dry branches, and NWFP free of cost Timber: State financed scheme: 25%; other finance 80%
Haryana	General Body on basis of majority vote • Final decision with CF but HRMS can seek compensation from Commissioner	All NWFPs Commercial produce lease to HRMS other income to be shared with HRMS 25 to 50% for timber
Himachal Pradesh	Individual membership can be dissolved by General Body; DFO can dissolve executive body • Final decision with CF	All NWFPs 25% of net sale proceeds of final harvest to village development fund
Jammu & Kashmir	DFO or RFO acting on behalf of DFO can terminate individual membership • Final decision with CF	All NWFPs 25% of net revenue from final harvest in cash/kind
Karnataka	RFO can terminate individual membership; DFO can terminate agreement without compensation for work undertaken • Final decision with PCCF	All dry leaves, lops, tops, grasses free of cost 25% of net sale proceeds to individual members; 25% to village forest development fund

Table 5.3 (continued)

State	Power to dissolve local organisation & final deciding authority	Benefit-sharing NWFPs & Timber
Madhya Pradesh	DFO can terminate the agreement • Final decision with CF	All NWFPs except for nationalised NWFP - share is 30% of net income 100% of fuelwood, poles, bamboos from thinning, 30% of final harvest of equivalent net income
Maharashtra	DFO can terminate the agreement • Final decision with CF	All NWFPs free of cost except cashew and tendu Final produce at max 50% market rate or 50% net income from sale (varies according to scheme)
Orissa	DFO can dissolve the executive committee • Final decision with CF	All NWFPs except those leased by FD, all intermediate products 50% of major harvest or 50% of net sale proceeds
Punjab	Not specified • Not specified	All NWFPs for own consumption. Fodder grasses to be auctioned at low rates to local people All revenue from private and community forests go to owners
Rajasthan	DFO can terminate agreement • Final decision with CCF & Director of Social Forestry	Free grass, fodder and other NWFPs (not bamboo) 60% of net sale proceeds
Tripura	DFO can terminate agreement on recommendation of RFO • Final decision with CF	All NWFPs free of cost Produce for own consumption and 50% of net revenue from sale of remaining produce
West Bengal	DFO can terminate agreement • Final decision with CF	Cashew 25%; sal seed, kendu leaves on approved tariff, others free • 25% of net income

Key to abbreviations
BC Backward Class; DFO District Forest Officer; FD Forest Department; FG Forest Guard; FLC Forest Labour Cooperative; GP Gram Panchayat; HRMS Hill Resource Management Society; MM Mahila Mandal (women's group); NGO Non Government Organisation; PS Panchayat Samiti; RFO Range Officer; SC Scheduled Caste; ST Scheduled Tribe; VLW Village Level Worker; YM Yuvak Mandal (youth group)

Sources: Sarin, 1993; SPWD, 1993; Femconsult, 1995a,b,c

- in cases where the value of labour involved in protection cannot be immediately mobilised because of the lack of intermediate products, other means of rewarding protection need to be put in place

- the percentage share of benefits between the government and local groups should be scrutinised and should bear some relationship to the value of the labour involved in protecting and managing the resource (Kolavalli, 1995)

In some of the most successful groups loans have been advanced to members who need money, and developmental activities in the village taken up, such as building a school and constructing a drinking water system (Pandey, 1994). Such hopeful signs indicate that some groups are vigorous and developing beyond the initial forestry entry point to encompass all aspects of rural life.

Future issues that are going to affect the sustainability and incentive for involvement in forest protection include the transferability of rights. Datta (1995) puts forward an interesting argument that rights should not be automatically transferable but should be associated with some minimal level of responsibility. If these conditions are not fulfilled then right-holders should be forced to leave the group. Similarly, those who exercise their rights with responsibility, i.e. they withhold use of the resource for future benefits, should not lose access to this resource if they move away from the area or if the benefits are to be received by a future generation. Datta suggests that there should be a 'periodic encashment of accumulated forest protection credits' to enable current generations to reap some benefits from their labour input. This notion of a futures market in potential values is particularly important in areas of poor resources where large labour inputs are required to develop the resource for minimal returns.

In terms of benefits to the state and achieving national objectives of improvement of the forest estate, it is apparent that forest resources are upgrading in areas where there is effective participation (Chatterjee, 1994; Femconsult, 1995 a,b,c). Again the question arises of whether the current sharing arrangements are fair, given the risk undertaken by local people and the opportunity cost of their labour as compared with the Forest Department which is effectively transferring the costs of protection and receiving large financial compensation in return (Poffenberger and Singh, 1992; Sarin, 1993; Kolavalli, 1995).

5.7 Discussion

The conclusions of a recent study (Femconsult, 1995a) that analysed the incentives for involvement in joint forest management are relevant to the discussion of this chapter and are reproduced in full:

- JFM provides a means whereby the state transfers some of its rights and responsibilities to locally constituted forest protection committees. The overall benefits and incentives to support this approach are different for the state, the forest protection committee and specific stakeholders.

- The primary objective of the process is to introduce joint management in order that local perceptions are factored into decision-making. The underlying assumption is that there is a convergence between the private incentives of forest users and the national objective of maintaining forest resources and that people will protect the forest since they have a stake in the outputs.

- The analysis suggests that this assumption is valid at the level of the forest protection committee (FPC) and that forest protection and conservation will become more effective under JFM.

- Although the analysis suggests that there are overall benefits to JFM in economic terms, as a result of effective protection and in terms of revenue, the economic benefits are more secure, again, as a result of protection. However, for some sub-groups within the forest protection committee, the effect of better protection is to cut off access to their livelihoods. The extent to which the perceptions and values of these sub-groups are factored into decision-making depends on the functioning of the FPC. The study concluded that in large, heterogeneous FPCs, the poorer, less powerful groups were marginalised as a result of the institutionalisation of forest management through JFM.

- The inclusion of local perceptions requires the active involvement of people in planning, management and decision-making as well as in implementation. In practice, participation has been limited to protection activities and wage labour for crop establishment. As a result, JFM appeared to be similar to other forms of 'welfare forestry' and was often seen as just another funding scheme which, with people's participation, will enable the Forest Department to protect the forest more effectively. This is also reflected in the observation that JFM is associated with protection and planting, rather than with management and decision-making.

- The introduction of JFM has formalised the role of the community in protecting the forest. However, there is still concern that decision-making is biased towards timber and revenue production, since the opinions of socially weak sub-groups and women, whose primary interests may be for non-timber forest products, are not reflected in decision-making by the FPCs. The existing economic relationships between sub-groups within each FPC determine each sub-group's bargaining power. As a result of formalising arrangements, sub-groups which previously had only marginal interest in the forest now have a stake in the forest as a result of their share in revenue.

- There are positive incentives based on perceived benefits for both the Forest Departments and FPCs under JFM. However, the pre-existing contractual arrangements and uncertainties undoubtedly act as a disincentive for some FPCs.

- The underlying complexity of both biological and socio-economic relationships supports and reinforces the principles underlying JFM – that optimum resource management requires the active participation of local stakeholders within an overall regulatory framework.

In Chapter 6 we move away from the social organisational aspects of participatory forestry to consider the technical implications of meeting the diverse needs of stakeholder groups.

Notes

1. Ostrom (1992) provides a useful definition of a community as: '. . . a set of people (i) with some shared beliefs, including normative beliefs, and preferences, beyond those constituting their collective action problem, (ii) with a more-or-less stable set of members, (iii) who expect to continue interacting with one another for some time to come, and (iv) whose relations are direct (unmediated by third parties) and multiplex'.
2. For example, in the state of Haryana, villagers' harvest of **bhabhar** grass, an important source of income, declined as the regenerating forest eventually shaded out the grasses. (In this case the local people were more interested in the income available from the grass, than in potential tree products, whereas the Forest Department used the grass income as an incentive to encourage local protection of the resources to enable a 're-greening' of the wastelands with trees, i.e. both parties had entirely different and irreconcilable objectives for agreeing to be involved in joint management of these areas) (Wollenberg and Hobley, 1994).
3. See Fernandes and Menon (1987) for a useful and detailed discussion of the involvement of tribal women in the forest economy.
4. The examples provided draw on work with Brahmin/Chhetri households. It is accepted that the interpretations provided here may not apply equally to other caste or ethnic groups. In particular, there may be further differences observed between hill and Tarai groups.

6 The New Silviculture

India: Jeff Campbell and BMS Rathore[1]
Nepal: Peter Branney

> Forestry is not about trees, it is about people. And it is about trees only insofar as trees can serve the needs of people (Westoby, 1987)

6.1 Introduction

As was stressed in Chapter 1, over the last two decades of forestry practice much attention has been focused on the social, organisational and institutional relationships. However, as the needs of forest users have become more clearly articulated through participatory forest management, so it has thrown into relief the inadequacy of the technical solutions offered, and the need to move local people away from highly conservative protection systems to active technical management (Arnold and Campbell, 1986; Branney and Dev, 1993; Fisher, 1989). This chapter focuses on the 'trees' and tries to give substance to the relationship between trees and people.The two case studies presented in this chapter focus specifically on the mechanisms through which to address the currently often divergent paths of people and trees in forestry. The discussion will consider the consequences of the new silviculture for the Forest Departments entrusted with its development, and will ask what further skills are required to allow a fully creative silviculture to be developed.

Old practices based on uniform silvicultural prescriptions developed under colonial and revenue-oriented systems will need to be re-examined. To meet a diversity of stakeholders' needs, both JFM and community forestry are forcing a re-evaluation of the variety of goods and services to be produced by forest lands. These may include grasses, small diameter fuelwood, timber for house con-

Plough shares, Bihar, India

struction and farming implements, tree fodder and the whole range of non-timber forest products. Since local stakeholders' livelihoods are still closely linked to forest lands this new multiplicity of outputs may increasingly have to come from the same forest areas. Managing for mixed species forests, which produce a variety of products and usufructs, requires the evolution of new management planning methods and the application in the field of innovative silvicultural strategies (SPWD, 1992). This multi-faceted form of management constitutes one of the most fundamental challenges the forestry sector has yet experienced[2] (see Table 6.1 which indicates the types of changes that need to occur).

6.2 The New Silviculture in India

Experience in India is now focusing debate on how to develop appropriate silvicultural systems to meet the needs of multi-resource management (Behan, 1990; Poffenberger and McGean, 1994). The following sections consider some of the lessons learned to date and the implications for future interventions. A number of stimulating presentations, papers and articles circulating in Indian forestry circles are beginning to discuss the evolving relationship between silvicultural issues and JFM approaches (Campbell, 1993; Chaturvedi, 1992a, 1992b; Guhathakurta, 1992a; Shah, 1994a, 1994b; Pathan, 1994; Maithani, 1994; Lal, 1994; Saxena, 1994; Rathore, 1994 Singh *et al.*, 1993). Drawing on these and on discussion with field foresters and community members, a number of components of silvicultural innovation for JFM emerge (see Box 6.1).

The social and institutional aspects of these approaches have been discussed in previous chapters; the focus here will be on the technical issues, and an examination of some of the innovatory practices being carried out in different parts of India. The case presented in Box 6.2 and Figure 6.1 illustrates the ability of local people to design complex management systems that fulfil a variety of forest users' needs, and contradicts much currently held belief that local people are not able to manage complexity without professional training.

Drawing on this approach, recent microplanning exercises in the Malpone Village of the Handia Range of the Harda Forest Division, Madhya Pradesh resulted in an innovative management plan which focused on a nested set of silvicultural prescriptions, dividing the forest and its management by species and canopy tiers. The forest is classified as a Southern Dry Deciduous Miscellaneous Forest. Under the management plan the forest under the village committee was divided into four blocks. The mix of silvicultural practices included species-based prescriptions, laying out particular plant

Table 6.1 Technology and forest management matrix

Issues	What works/can work	Changes needed
Technology and management		
1. depend on people's needs and expectations:		
• assessment of indigenous technologies and management systems	PRA/RRA techniques, ground observations, use of secondary data, historical data	training for forest officers, field staff in methodologies
• screening of new technologies – individual species level – multiple species mix – silvicultural practices – area management	literature review, demonstration, analysis of packages	user orientation which is location-specific, flexible, promotes diversity, and offers both early and regular benefits
2. Rapid research and experimentation		development of methodology, identification of appropriate persons, field networking
3. Incorporating economic analysis	PRA/RRA techniques	studies of yield and mensurational data, cost-benefit analysis, market analyses; use of valuation techniques, including environmental values
4. Training/education (of Forest Depts, local people, NGOs)	workshops, cross visits, field demonstrations	revise curriculum, offer follow-up training, encourage integration between departments, and between GO and NGO
5. Processing and marketing of forest resources	appropriate technologies, market linkages, low levels of capital inputs based on sustainable resources	encourage marketing of non-timber products, training in local value-added processing, bank credit and infrastructural support services
6. Forestation/natural forest management	regeneration	promotion of high-value, market-oriented products grown in both horizontal and vertical strata.

manipulation strategies for NTFPs such as **tendu** (*Diospyros melanoxylon*) and **mahua** (*Madhuca latifolia*), fodder trees and fuelwood plants. A locally abundant tree, *Writia tintoria*, was to be managed for fuel on a four-year felling cycle. Different thinning/harvesting schedules were prescribed at 4,8,12,16,20 and 24 years for poles and timber. Rotational grazing in the worked area for a period of four

Box 6.1 Principles of Innovative Silviculture and Management

- participatory objective setting
- participatory management prescriptions
- managing for multiple products
- multiple time horizons
- site-specific prescriptions
- landscape-level linkages
- maximisation of growing space
- encouraging natural regeneration
- mimicking natural forest in plantations
- more individual plant manipulation
- participatory implementation
- innovative grazing and fire control systems
- frequent impact assessment and monitoring
- flexible, decentralised planning and budgeting
- increased accountability
- equitable benefit-sharing

Box 6.2 Participatory Silvicultural Planning Practice in West Bengal

In West Bengal a participatory silvicultural planning exercise was undertaken with local people.

1. The degree of attention which is given to establishing the correct ratio of product availability over time was impressive. Participants constantly coach whoever is placing stones or assigning a number and revise the ratios until a mutual consent is reached. Men, women and different user groups within the community will have different lists of products and different priorities within products. This makes it imperative to undertake the exercise with different groups and both genders.
2. People quickly grasped the purpose of the exercise: 'This is the history of our forest . . .' 'This shows what we need to do to get back the good old forests.' In one session a long digression took place as one woman intervened: 'First we should discuss the reasons for these changes in the forest.' This led to an analysis of forest use in the area and the causes of forest degradation.
3. People had detailed knowledge about the ecological conditions necessary for different kinds of products. This came out in various discussions, particularly with relation to the soil and site conditions necessary for abundant mushroom production on the forest floor, or to the reasons why date palms were more abundant in degraded sites.
4. People were able to suggest a variety of approaches for the manipulation of single trees and plants, and for sequencing different operations, having closed periods at times of peak seedling regeneration, etc., but had not had the opportunity to consider management options for the forest as a whole, particularly related to sequencing of harvest operations for poles and timber. Some suggestions and concerns echoed technical issues, such as heart rot in older sal trees, which they had learned from department staff. When discussing ways to resolve the conflict between the need for fuelwood and cash (from short rotation pole wood harvests) and the need for mature timber, seed and fruit-bearing trees and biodiversity, the idea of rotations, and a sequential felling series, came to one elderly tribal man like a bolt from the blue.
5. In general, people finally arrived at a mixed model of an uneven-aged forest with zones for biodiversity protection (sacred grove); timber, seeds and mushroom production (high forest with selection felling and minimal use); a felling series of coupes of different ages (very short rotation for fuelwood and leaves, and longer rotations for poles); and special management for certain species – such as fruit trees or tendu bushes.
6. It was clear that this was the first time these people had really been consulted about the management of the forest as a whole; they were still unsure about whether their suggestions would amount to anything. Before undertaking this exercise Forest Department staff need to be committed to JFM and open to making silvicultural changes where technically feasible to meet village needs.

Source: Jeff Campbell, pers. comm.

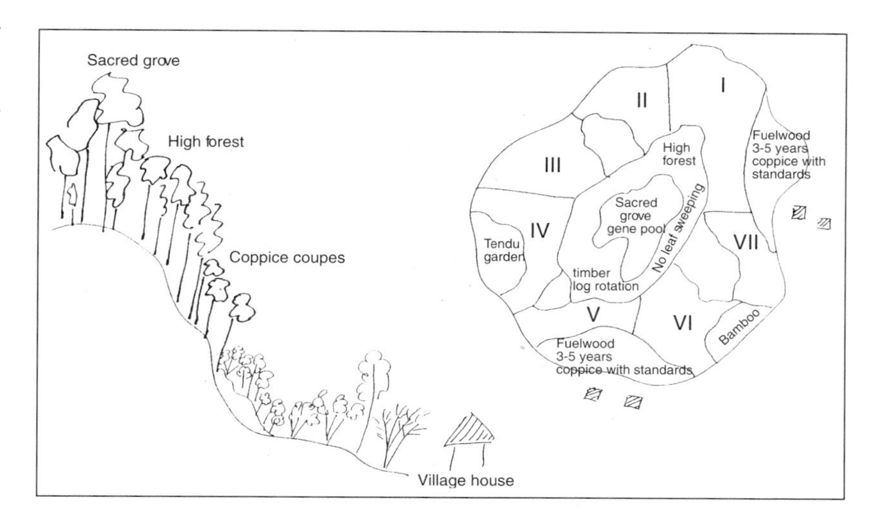

Figure 6.1
Silvicultural diversity

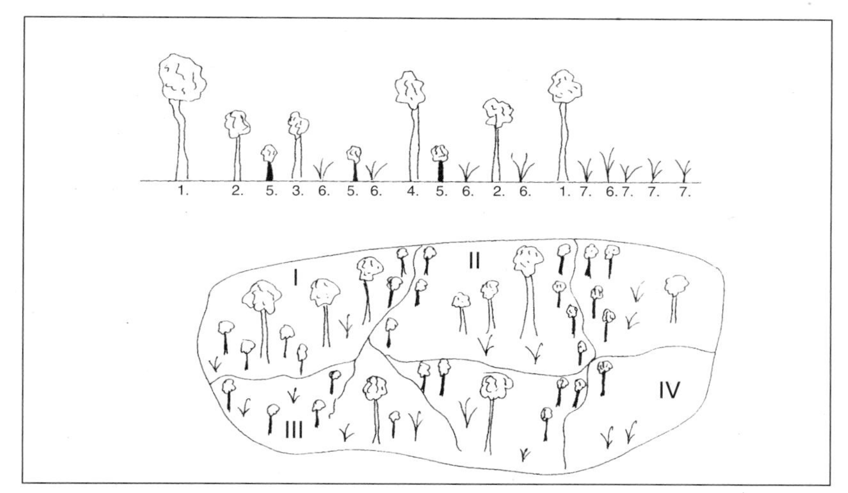

Figure 6.2
Malpone's Management Plan
Species and Management Index
1. Mahua
2. Lagerstroemia (stool singling at 4 year interval)
3. Tendu bushes (pruning as per established regime)
4. Tendu tree (not to be felled)
5. *Writia tintoria* (to be worked on 4 year felling cycle)
6. Bamboo (planted to be worked on 4–4 year cycle after establishment)
7. Grass legume (to be managed on cut and carry basis with canopy manipulation to allow increased light to grass level)

years was also prescribed. Figure 6.2 illustrates this plan.

This village-based exercise was combined with physical assessment of the resources by foresters and local people. In Harda, this was done by laying out and harvesting a number of 0.1 hectare sample plots to assess regeneration and yields of fodder, fuel and small timber. In Gujarat, the Aga Khan Rural Support Programme has developed the participatory root stock transect exercise, in which local people count all the existing root stock that has regenerative capacity following protection.

Managing for multiple products

The multiplicity of stakeholders leads to a multiplicity of management objectives. A number of important questions need to be answered in order to tailor silvicultural approaches for multiple use:

- How does one experiment and evaluate with a mix of silvicultural practices?

- How can one maximise fodder, fuelwood and a range of NTFPs as well as timber from a small patch of forest and turn out the optimal mix of economic flows?

- Can a steady supply of fuelwood and fodder be produced for local use?

- Is it possible to sustain a certain amount of fuelwood cutting for sale as well, and still allow other forest management objectives to be addressed?

- What level of grazing intensity is acceptable in different stands?

- If the users' objective is the production of grass, why grow trees as well that later shade out the grasses (Panda *et al.*, 1992)?

- What is the impact on the ecological character of the forest if extraction rates are intensified for different products?

These questions are the ones currently under discussion in India, much of the debate being highly controversial. An example of the controversy surrounding the silvicultural practices of the sal forests of South West Bengal is illustrated in Box 6.3.

During a visit to forest committees in Bankura North Forest Division, discussions were held with a number of villagers. They

Box 6.3 Is it Coppice with Standards or Conversion to High Forest?

The recent exchanges in *Wasteland News* between Dr Chaturvedi of the Tata Energy Research Institute and foresters from West Bengal, including Mr Guhathakurta, retired PCCF, about the most appropriate silvicultural system for the sal (*Shorea robusta*) forests of South West Bengal have raised some important questions (Chaturvedi, 1992a; and 1992b; Guhathakurta, 1992a). However, the discussion needs to go beyond contesting macro-level prescriptions (e.g. coppice with standards versus conversion to high forest) to an approach that is site-specific and flexible. For example, in the Arabari case, it may no longer be possible to convert the whole forest from a coppice with standards system to long rotation cycle dependent on advanced regeneration, which Chaturvedi advocates for ensuring the long-term genetic viability of the forest type. However, it may be possible to set aside certain good quality forest patches at the village forest committee level and such patches could be worked on a selection system. Similarly it may not be practical to stop completely the sweeping of sal leaves from the forest floor, which as Chaturvedi rightly argues, interrupts the nutrient cycle and impoverishes the forest floor, in the entire sal belt. However, it may be possible to reduce leaf sweeping, and perhaps stop it completely in certain sections of the forest.

Sal regeneration, West Bengal, India

were asked about the ecological effects of sweeping leaves from the forest floor and admitted to the adverse impact the practice would have on the forest. At the same time, they raised the question of how to minimise leaf sweeping, as leaves are an important source of fuel, especially for parboiling rice. Discussions centred around site-specific planning indicated that it is possible to meet the leaf needs of the villagers partially, whilst maintaining the ecological integrity of the forest floor.

In addition, there is a valid concern that biodiversity may not necessarily be increasing with community protection: that the forests are becoming, if anything, even more purely sal monocultures than under previous silvicultural regimes. This is partly due to the local classifications of trees as **kaath** (good timber) and **akaath** or **kukaath** (poor timber), with most communities enforcing felling bans only on **kaath** species which in this area primarily means sal. This may be resulting in selective pruning of all **akaath** species, relegating them permanently to the shrub class with even less chance of becoming seed bearers than the coppiced sal (Bhattacharya, 1994).

Site-specific planning

Just as the diversity and complexity of social systems have been recognised, so silvicultural responses must be equally site-specific and responsive to local conditions. Recent work in Rajasthan indicates that it is possible to create management-objective zones within a community-protected forest. The following guidance areas have been developed, based on experience from different parts of India. They indicate the types of questions that have to be considered when working with local people to decide how a site may best be managed, and to develop silvicultural responses appropriate to different conditions.

Encouraging natural regeneration

To ensure that the growing space is maximised it is essential that all levels of the forest architecture should be used. This should include use of shade-tolerant understorey trees and shrubs; management of shrub and herb layers; introduction of herbaceous medicinal plants; management of forest floor to enrich soil and encourage natural regeneration and the production of edible tubers. If JFM is to be replicated widely, low-cost options based on natural regeneration are more viable than strategies based solely on plantations. However, to assess whether a site is capable of supporting natural regeneration, a series of nested questions need to be asked to decide what level of intervention will be required to ensure the success of the natural regeneration option. A 'decision-tree' approach is helpful as it allows a series of options to be assessed. To give an example:

- Is there adequate root stock or advanced growth to ensure natural regeneration without any inputs other than protection?

- If not, is there sufficient root stock density and/or advanced regeneration for a minimum of soil working and site improvement with some enrichment planting or gap filling to be enough to supplement and enhance the natural regeneration?

- If plantation is necessary because root stocks and advance regeneration are rare, is it possible to plant to ensure that existing root stocks are encouraged to grow as part of the mixed stand? For example, in a number of areas in southern Rajasthan and northern Gujarat where fully stocked plantations were undertaken on supposedly barren lands, teak coppice growth from semi-buried and quiescent stumps emerged after several years of protection (see Box 6.4 for a similar example from Nepal). Experience from Madhya Pradesh indicates that deferring the planting activity for 3–4 years on barren lands, while the closed area is treated with soil/moisture conservation inputs, allows the regeneration of grasses and assessment of the viability of root stocks.

Chir pine plantation, lopped for firewood, Nepal

Favouring natural regeneration does not mean that the plantation option has been rejected, but just as there is a continuum of institutional arrangements, similarly there is a continuum of silvicultural options. The continuum approach means that the focus shifts to enhancing the growth potential of existing plants and emphasising local diversity. There will always be a series of inter-

Box 6.4 From Monoculture to Species Rich Natural Forests: The Experience of the NACFP

One of the most successful plantation species is the indigenous chir pine (*Pinus roxburghii*). It is a tough, hardy pioneer species that occurs naturally at elevations around 1300 m. Many of the potential plantation areas are heavily grazed, often eroding grasslands with shallow stony soils. From an ecological point of view chir pine is one of the few species which can survive and grow on many of these sites. It is also easy to handle in the nursery, making it well suited for use in small village nurseries. Experience has shown that attempts to grow more desirable broadleaf species in large-scale plantations have largely failed. This is particularly true in the drier parts of central Nepal, although some of the better watered areas have had some success with moisture-loving species such as utis (*Alnus nepalensis*).

If a plantation area receives protection from grazing, a range of tree and shrub species often invades the site. This is most evident on the moister northern aspects at elevations above about 1400m. Conventional forestry would have us remove any competing 'weed' species that were adversely affecting the growth of the chir pine. However, many of the invading species can provide the product mix needed by farmers, so they need to be brought into the stand as a valuable part of the forest even if this means sacrificing some increment on the planted trees. The ecological role of chir pine is to act as the pioneer in returning the site to forest which can then be manipulated silviculturally to provide the products needed by the farmers.

A forest where regeneration was studied provides an interesting example of how species diversity can change from an unpromising beginning.

In a plantation forest 40 km north-east of Kathmandu there have been three waves of regeneration following pine planting. The first wave developed as coppice from stumps which were remnants of the original forest. The protection from regular cutting which accompanied pine planting allowed coppice shoots to survive and these became an early component of the stand, along with the pine trees. Most of these coppice shoots were *Schima wallichii*, a widely occurring broadleaf which produces highly valued firewood, construction timber and leaves for animal bedding. The second wave consisted of seedling regeneration which germinated about five years after plantation establishment. Like the first wave, the dominant species in this second wave was *S.wallichii*. At five years, the plantation would have been quite open. However, sufficient amelioration of the site must have occurred to provide a suitable seedbed for *Schima*. A dramatic change in species composition began at about 12 years, when the canopy had closed. A third wave of regeneration occurred at this time, and it included a large number of useful fuel and fodder species. The most notable of the newcomers was *Litsea polyantha* which is highly valued for its leaf fodder. It regenerated at high densities and by plantation age 14 years it covered the forest floor at a density of about 1600 per ha. Other valuable species which appeared at the same time included *Fraxinus floribunda, Cedrela toona, Castanopsis indica*, *Prunus cerasoides* and *Michelia champaca*.

Source: Gilmour and Fisher, 1991

ventions that run from total dependence on natural regeneration to total dependence on plantations. Innovative approaches can extend the boundaries of the use of natural regeneration and allow experimentation in areas of overlap. Much still needs to be understood about these different approaches; Box 6.5 indicates the range of action research questions still to be addressed.

If we accept the importance of relying more heavily on natural regeneration because it meets the multiple objectives of users and is more cost-effective, it underlines the argument that in general natural forests are more appropriate for meeting local people's needs. However, where forests have become so degraded that it is not possible to rely on natural regeneration plantations will have to be undertaken. Plantation strategies have already moved away from the monocultures of the social forestry era (Shiva *et al.*, 1982) to more complex polycultures with a variety of different species. In Gujarat, plantations may include as many as 21 species of so-called 'multi-purpose' trees. In addition, a number of foresters are arguing forcefully for both multi-species and multi-tier plantations that mimic more closely the natural variety of overstorey and understorey plants (Guhathakurta, 1992b; Pathan, 1994). In Madhya Pradesh and West Bengal a number of new silvicultural models

Box 6.5 Outstanding Questions about Forest Regeneration

Once protection is in place

- What is the pattern of forest regeneration?
- Which species predominate?
- Which species are suppressed?
- What are the livelihood implications if there is differential survival of species?
- What changes in succession are likely over what period of time?
- What is the rate of regeneration for different species?
- What is the total increase in biomass?
- How does the natural regenerative process change under various types of management (open or rotational grazing) or different silvicultural manipulation (i.e. cleaning, stool levelling, multiple shoot cutting)?
- In different regenerating forest ecosystems, what are the yields and the rate of increase or decrease in the productivity of important NWFPs? (These might include seeds, fruits, flowers, mushrooms, fibres, fodder grasses, fuelwood, etc.)
- What effects do various enrichment planting strategies have on the volume yields of important commodities?
- What levels could be achieved for a maximum sustainable yield?
- What is the total value of produce for each year of regeneration?
- How do the total productivity, value and flow of products and income change with different forest management practices (i.e. closure to grazing and fire, cleaning, multiple shoot cutting, enrichment planting, spatial designs, etc.) and different harvesting strategies (lopping, pollarding, coupe-cutting systems, rotation ages)?
- What is the cost of inputs at various points during the regeneration process?

Silviculture options should be based on a clear assessment of client needs and of the ecological capacity of the site to support these needs.

have been introduced for plantations to begin to mimic successional patterns of natural forests.

Once again, the interface between technical and social innovation must be carefully managed, and the composition and management of these plantations should be negotiated on a site-specific basis with the users of the land. Simply applying more complicated models, and responding to the site's ecological potential, does not mean that joint management has been fully embraced. As the variety of tree, shrub and grass species increases, innovative planting practices will also need to be evolved. For example, many species propagate well by direct seeding and may not need to be raised to the sapling stage in nurseries before planting out.

Multiple time horizons

Many forest products are seasonal in nature – fruits, fresh leaves and flowers – while others, like certain fodder or fibre grasses, can be harvested periodically. Fuelwood often consists of twigs and branches which can be harvested on a periodic basis, every few years, while poles are of medium duration and timber requires very long temporal rotations. People depend on a steady supply of different products from the forests over a multiple time horizon. The new silvicultural approaches will also need to accommodate these temporal sequences and events with a more complex series of silvicultural activities over both time and space than is currently advocated through occasional thinning regimes under most standard silvicultural prescriptions. They will also require an understanding of the impacts over time of different activities on different livelihood-dependent users. The value of economic flows from the forest, aggregated over seasons and years, should not be underestimated. Trading some of the delayed returns from timber in favour of more frequent shorter-term returns may make good

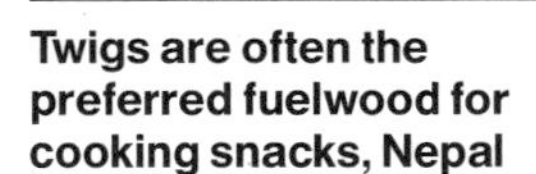

Twigs are often the preferred fuelwood for cooking snacks, Nepal

financial sense and fulfil the needs of a larger number of people.

Individual plant manipulation

Indigenous practices of arboriculture and vegetation management include a wide variety of individual plant manipulation methods including pruning, lopping, pollarding, ratooning, weeding and protecting stems and fruits. In the same way, modern plantation management includes increased attention to micro-site amelioration and soil and moisture conservation at an individual seedling level. How can these cultural practices be applied more intensively within forests themselves to help meet multiple objectives? Pathan (1994) outlines several strategies for drastic thinning to open space for fruit tree crowns to develop, with the intention of creating a forest 'orchard' of about 250 NTFP-yielding trees per hectare. While contesting the theory that all trees in JFM systems should be managed purely for crown-based products (Saxena, 1994), there is much to be said for increased attention to managing the productivity of foliage, small stems, fruits, etc., through more frequent pruning and pollarding, as one important component of the silvicultural mix. If such individual manipulation for one type of product is carried out, the implications for other products must be considered, since such manipulation may reduce stem and timber production. An example of these approaches is provided by Sabarkantha Division in Gujarat, where the DFO, H.S. Singh, has encouraged members of the Tree Grower's Co-operative Society to stimulate root suckers of **tendu** shrubs, resulting in a rapid increase of foliage on this economically important plant, and the creation of a 'tea plantation' effect of tendu in one stand of joint managed forest.

Innovative grazing and fire control systems

Recent experiences from Harda Forest Division in Madhya Pradesh indicate that there are some potential solutions to both these consuming problems of forest management. The Forest Department has encouraged JFM groups in the protection and management of both degraded and fairly well stocked, older teak forests. Although JFM is still regarded by many as a strategy for regenerating forests which are already heavily degraded, Harda's experience illustrates the importance of people's involvement in the protection and management of better quality forests.

Thousands of hectares of forests in Harda Division have experienced gregarious flowering of bamboo in the recent past, an event which occurs every 35–50 years depending on the species of bamboo and other edapho-climatic factors. Without protection from annual fires and heavy grazing pressures, bamboo regeneration after flowering is almost impossible and the subsequent death of flowered clumps inevitable. Encouraging results have been obtained in a large area where bamboo flowered in 1986. Regenerating bamboo was on the verge of extinction due to localised grazing pressure and recurrent forest fires. Having put the area under participatory grazing and fire management since 1990–91, the local forest protection committee, in partnership with the Forest Department, have now ensured a significant recovery of bamboo. These management practices have also had an additional benefit in the regeneration of a multiplicity of other species. In addition, there has been a substantial production of grasses in areas that were hitherto considered to be unable to sustain grass production. Two adjoining forest protection committees were able to earn more than Rs 100,000 from grass sales in the 1993 season.

Fire protection has also been undertaken by forest protection committees which have developed their own fire lines and back-firing strategies based on a PRA exercise in which they analysed their own use patterns. Innovations include back-firing along all footpaths, and not just major fire lines, as people suggest that this is where past fires have started (due to dropping **bidis** (local cigarettes)). Local groups have also agreed to stop setting annual fires under **mahua** trees, a common practice in central India, undertaken to facilitate flower collection. They claim that their effectiveness in collecting fallen flowers has not been reduced. In fact, they say that flower collection has become even more difficult for their major competitors – the monkeys.

Participatory implementation

Many of the components of innovative management and silvicultural practices discussed so far call for an intensification of activities and increasing operational complexity. By extending the participation of local people from decision-making to implementation, the range of new practices discussed can become operational without substantial direct costs to the governmental organisations. However, as has been discussed in Chapter 4, there are significant additional costs differentially incurred by local people. There are several questions to be addressed to allow a smooth transition from joint decision-making to joint implementation: how can local people become equally involved in planting, thinning, pruning, harvesting, processing and trade? What new budgeting procedures will need to be developed to channel funds for different activities through the forest management and protection committees directly? What processing technologies and marketing linkages for trade in forest products will have to be forged to increase involvement in value-addition possibilities?

A second example from Harda illustrates people's involvement in management decisions in a degraded forest area, which had been protected by local groups for three years. Forest staff and local labourers were involved in a multiple shoot cutting operation (MSC), in which a clump of teak coppice shoots were thinned out, leaving only one or two of the straightest and tallest shoots growing. The DFO asked the forest staff what the benefit-sharing arrangement was for the cut shoots, which are an excellent source of fuelwood. On hearing that it had not been decided before the operation was carried out, the DFO immediately convened a village meeting and instructed his own staff and the local people that no forestry operations were to be carried out unless benefit-sharing mechanisms for the products were clearly agreed. This resulted in a discussion in which it was agreed that people would be allowed to take away as many headloads as required per household, but that there would be a moderate fee for cartloads and not more than one cartload per household could be taken. Once the discussion began, several villagers mentioned that the MSC operations were being carried out very roughly, and that the same pruning procedures used for teak were being applied to other trees, including fruit trees. This revealed the need for different cultural operations, varying from species to species.

The third and most interesting management issue is centred around the plucking of **tendu** leaves. These leaves are used to roll **bidis**, a low-cost cigarette, widely smoked in India. Tendu leaves constitute the most important NTFP in Madhya Pradesh. Government monopoly of the collection and sale of raw leaves brings in up to Rs 3 billion to the State Treasury annually. Local people complained that the benefits from tendu collection are not proportionately as substantial as they were in the days of the contractors, before the State Government turned the trade into a monopoly controlled by state-run co-operative societies. There are several reasons given for this reduction. First is that thorough pruning of tendu shrubs and trees is no longer possible, because the government pays only a small amount (Rs 5 per standard bag collected) to the village labourers to undertake pruning. As a result only a small proportion of the trees are pruned – and most of the bushes in the forest do not produce the new flush of good quality leaves without adequate pruning. According to villagers, when a contractor purchased the lease for a given area, it was in his interest to see that as many bushes were pruned as possible in order to maximise the yield of high quality leaves. Second, the collection period has been reduced considerably as the government has a pre-designated quota for each collection unit, and needs to stop all collection well before the onset of the monsoon to ensure that all the leaves, collected in standard bags, are safely transported before the rains. This means that the availability of leaves far exceeds the amount actually collected. Again, in the days of the contractor the collection period was extended right up to the last day before the monsoon since the objective was to maximise return, and move stock efficiently and regularly. Although the co-operatives have helped to set higher rates per standard bag, the number of working days are fewer, and the yield is reduced.

At Harda, the DFO who had worked with one village forest committee in a tendu collecting area decided that this was an area in which JFM could be extended to include partnerships for the collection and sale of tendu leaves. The first proposal put to the village committee was that it should be given sole rights over tendu collection. This was suggested in response to villager complaints about labourers from outside the village collecting tendu leaves.

Villagers agreed to this but suggested that they would need to extend the number of collecting days, since there would now be fewer collectors. Women members of the village committee also responded well to this suggestion and related how outsiders were ruthless in their collection techniques, stripping branches of leaves regardless of their quality, whereas collectors from the village, who were able to return to the same plant several times, selectively picked only ripe leaves. The DFO also suggested that the village should invest in large-scale pruning of the trees and bushes to enhance yields, and promised that the Forest Department in turn would try to increase the quota and extend the collection period.

This is an interesting experiment where the local people's relationship with the Forest Department has changed from one of wage labour to one of shareholder in the forest resource. With this changed relationship comes responsibility and investment in the resource prior to extraction of any benefits. In this case pruning would have to be carried out by local people without payment, on the presumption that enhanced future benefits would compensate for the early risk. As has been discussed in Chapters 4 and 5, for poorer people whose livelihoods are often dependent on forest-based cash employment the removal of such sources of income through new JFM arrangements may jeopardise their livelihoods. In such cases, it has been suggested that forest committees could sub-contract many of these activities to poorer village people and pay them from the budgets previously held by Forest Department staff. Questions of accountability still remain to be answered, as do those concerning mechanisms for ensuring the flow of funds to the village level.

The choice of silvicultural system has a direct impact on local people whose livelihoods are dependent on access to forest products.

6.3 User Groups as Forest Managers: the Experience of the NUKCFP

Handover of national forest to communities and formation of forest user groups are not intended to be ends in themselves, rather the means to an end: namely to ensure that rural communities can meet their present and future needs for forest products in a sustainable way. The creation of a category of forest owned, utilised and protected by local communities has now been recognised as being the only realistic means of halting the slow degradation of forests in the Middle Hills of Nepal. Given the aim of securing forest products in perpetuity, the success of community forestry cannot be measured simply by the number of forest user groups formed nor by the area of forest handed over to them, but rather by the effectiveness of these groups as forest managers, developing and utilising their forest resources to meet their own specific requirements.

The important challenge still outstanding is for established user groups to demonstrate that they have the ability to manage their forests by themselves, and for Forest Department field staff to show that they can act as effective extension agents providing the type of assistance user groups need in order to achieve this. The NUKCFP is paying increasing attention to the management aspects of post-formation support to forest user groups by developing the concept of participatory forest management. This involves user groups, assisted by Forest Department field staff, in preparing and implementing their own working plans for community forests which try to incorporate the diverse needs of all users as well as improving the forest's future capacity to supply forest products.

The need for effective forest management

Very little information is available concerning actual forest management by user groups beyond straightforward forest protection mechanisms which have been described in Chapter 4.

Too often in Nepal good forest management has been equated with complete forest protection. In fact, this was the whole basis for the largely unsuccessful policing and protection role which until recently the Forest Department attempted to maintain throughout the country. Even where forests have been managed under indigenous protection systems for several decades, they may now be in good condition but their potential productive capacity is not being fully realised, since local people have little access to the forest products they need. Frequently, such forests have been allowed to become overmature and sometimes their condition may actually be deteriorating.

Newly formed forest user groups also tend to follow this pattern, adopting conservative and protectionist management strategies. Immediately after their formation, most user groups implement protection measures, allowing only collection of dry wood and twigs for fuel as well as certain non-wood products such as leaf litter for animal bedding and compost. To all intents and purposes such forest is closed to any productive forest management. Under certain circumstances where forests handed over are severely degraded, or consist mainly of young plantations, this approach is logical in terms of building up the resource base before initiating more active management. In other situations strict protection may reflect the fragility of the user organisation, in that it is easier to forbid general access to a forest than it is to control regular harvesting of products, where one group may exploit their right to use the forest at the expense of other members of the organisation.

Many user groups already control significant areas of valuable and potentially productive forest (especially natural forest) often in good or visibly improving condition. Contrary to the concerns of some people (see Box 7.3), there is no evidence to indicate that they are exploiting this change of ownership and causing increased forest degradation. In fact, experience shows that almost the opposite seems to be taking place. Such forests have the potential to

Protection of community forests leads to increased use of non-protected forest areas. There is a need to cluster support for the development of user groups in the same area, to ensure that all areas of forest come under active management.

supply a significant proportion of the community's own forest product needs and, in some cases, a surplus for sale outside the user group. Due to an emphasis on protective forest management, this potential is often not being realised.

As more forests are handed over to user groups as community forest, the result is that less national forest is available to local people for what has been effectively unrestricted collection of forest products. This is now becoming increasingly apparent to both Forest Department staff and established forest user groups. Both are

A well protected area of natural forest, Nepal

starting to recognise that more productive use needs to be made of community forests to satisfy forest product needs.

Studies in locations where most of the accessible forest has already been handed over to user groups, such as Bokkim village development committee (VDC) near Bhojpur bazaar, indicate that people are being forced to make more use of the tree and forest resources on their private land, thus disadvantaging poorer households who own less land. Furthermore, remaining pockets of national forest appear to be suffering from increased exploitation and illegal cutting activity. This leads to appeals from local people for the DFO to also hand these areas over to user groups (Young, 1994; Maharjan, 1994). In addition, firewood cutters, traditionally amongst the poorest members of the community and often owning little or no land of their own, are obliged to carry backloads of firewood from much further afield for sale in the bazaar, or else are forced out of business altogether. Without alternative sources of supply for essential forest products, protectionist attitudes by user groups cannot be sustained in the longer term.

Clearly, handover of forest without subsequent support to ensure that user groups become more effective forest managers can have a number of negative and undesirable effects. Unless the causes of these can be tackled, rural people, especially the poor who potentially have most to gain from community forests, will see the whole programme as yet another barrier to lifting themselves out of the cycle of poverty and deprivation. Without the popular perception amongst all members of a forest user group that they have something positive to gain, their participation will be minimal. As a result, the user group concept as an effective community-based organisation becomes less valid.

Developing the role of DFO field staff

How effective are District Forest Office field staff in providing support and advice to established forest user groups to enable them to undertake more effective forest management? Trained rangers and forest guards can draw on various sources of information when undertaking their advisory role, including:

- Technical training
- Personal working experience
- Published information
- Experiences of others, especially forest users

Forestry technical training has in many cases been recently supplemented by reorientation workshops aimed at enhancing the skills of field staff allowing them to become community forestry extension workers. However, there is often still a poor understanding of how previously learned technical skills relate to the present community forestry environment in which user groups themselves need to become forest managers. This problem is compounded when previous formal training has been forgotten, or when it tends to emphasise less appropriate aspects such as even-aged plantation forestry, rather than irregular natural forest management systems.

The personal working experience of field staff has also often been somewhat limited. Although many field staff may now have had considerable experience of using participatory approaches to forming forest user groups, they may never have been involved in any practical forest management work apart from plantation establishment. Frequently they have had no experience of making practical management decisions (such as marking thinnings or fellings), based on management objectives and actual forest conditions. Only rarely have they had experience of preparing and implementing forest working plans. Clearly an improvement in their practical abilities is required if they are to act as effective sources of advice and support to forest user groups.

Any published material directly relevant to participatory forest management by user groups is unlikely to be found at range posts in remote Middle Hill districts. It is even less likely to be written in Nepali – the main language of rangers and forest guards.

The knowledge and experiences of forest users who know the community forest well are likely to be of vital importance for effective forest management. However, field staff may not recognise the relevance of such indigenous knowledge, and may not be skilled in seeking it out and using it in a way which will assist in the management process.

As key implementors of the community forestry programme, Forest Department field staff are therefore not yet able to provide the type of support user groups need to enable them to make the shift from forest protection to effective forest management. Consequently, the approach of NUKCFP has been to provide assistance in four key areas:

- Developing technical forest management knowledge amongst field staff in those areas most relevant to forest management by user groups

- Providing opportunities for field staff (and user groups) to gain practical experience in forest management

- Providing written guidelines and information (in Nepali) relevant to user group forest management

- Assisting field staff in recognising and acquiring the techniques needed to incorporate users' own knowledge into their forest management.

Participatory forest management processes

The approved Forest Rules, passed in 1995, relating to the 1993 Forest Act recognise that two distinct stages need to be completed before a forest can be officially handed over to a forest user group by the District Forest Officer. First, the user group associated with a particular forest area needs to be formed and registered. This means that individual users (households) who are to be members of the group need to be identified; rules, regulations and administrative procedures relating to use of the forest have to be agreed by all users; a representative committee has to be chosen; individual responsibilities have to be assigned; and the community forest boundaries have to be clearly described or mapped. This leads to the production of a written user group constitution or **bidhan** for submission to the DFO.

The second stage is for the user group to produce a forest working plan, or **ban karyojana**, approved by an assembly of all the users, and also by the DFO (see Box 6.6). This contains more detailed information about the management aims of the group; a detailed forest description; and a description of harvesting and other forest management activities which the group intends to carry out. Once these two stages are complete, the forest is handed over and the user group receives a certificate of ownership, or **praman patra**, from the DFO.

From the point of view of the field staff actually facilitating the handover, this two-stage process can be a lengthy one. Since there is considerable pressure on field staff both from the DFO and from forest users to hand over forests as rapidly as possible, there has been a tendency to try to contract the two-stage process into a single one. In practice this has not been successful since it then becomes necessary to discuss and reach consensus on the very broad range of issues contained in both the constitution and the working plan at one single assembly of all the forest users. Invariably discussion focuses on the constitution at this point, since it is of more immediate priority, with the result that management provisions are generally inadequately covered. Consequently, forest areas are then handed over to user groups with only minimal provision being made for effective management and not surprisingly, in the absence of encouragement to do otherwise, user groups tend to protect rather than manage.

Identification of users and development of management practices have to be carried out carefully since they have major impacts on the future development of the user group's capacity to manage the forest according to its needs.

To address this problem, a participatory process has been devised where the ranger works closely with the user group in assisting them to prepare a working plan for their forest. This can be done at the preliminary stages of user group formation, and before the formal handover. However, it has been found to be more effective with established user groups who have themselves realised that they need a useful working plan for their forest and have assigned their committee the task of preparing one. In this case the prepared plan is simply attached to the existing documentation of the user group following approval by the DFO.

The term 'working plan' is used to emphasise its characteristic features distinguishing it from traditional forest management plans. These are illustrated in Box 6.7. If these features are not incorporated, it is unlikely that a user group will implement its working plan.

Participatory working plan preparation

Preparation of the working plan involves a participatory process,

Box 6.6 Translated Section of the Constitution of Mainakhop Giddhekhop Community Forest User Group, Dhankuta District

Function, rights and duties of the user group committee

The committee will get the users to perform the activities mentioned in the constitution and the working plan, and will implement the rules and regulations. The committee cannot formulate any new rules itself – only the users' general assembly can formulate rules or amend existing rules. Such rules can only be enforced after they have been agreed by the user group and approved by the DFO.

- to hold regular committee meetings at least once a month to discuss issues concerning forest management, problems being faced and future problems likely to arise. The meeting will only be considered valid if at least 9 committee members attend.
- to make arrangements for extraction, distribution and use of forest products and to fix the times and turns of users.
- to fine and punish those who act against the provisions of the working plan, or who violate rules, and to deposit the fine in the user group fund.
- to fix the users' turns for forest protection, and to inform those users.
- to carry out extension and publicity about how to prevent and control forest fires.
- to liaise with the District Forest Office and make arrangements for necessary technical and other support.
- to call a general assembly of users, giving at least 7 days notice, when the need arises to make any amendments to the constitution or working plan, and to send any new rules to the DFO office for approval.
- to inform all users of decisions made at user group committee meetings.
- to prepare the forest working plan and to submit it to the users' assembly and the DFO office for approval, after it has been passed by the general assembly. . .

Box 6.7 Characteristics of a Forest User Group Working Plan

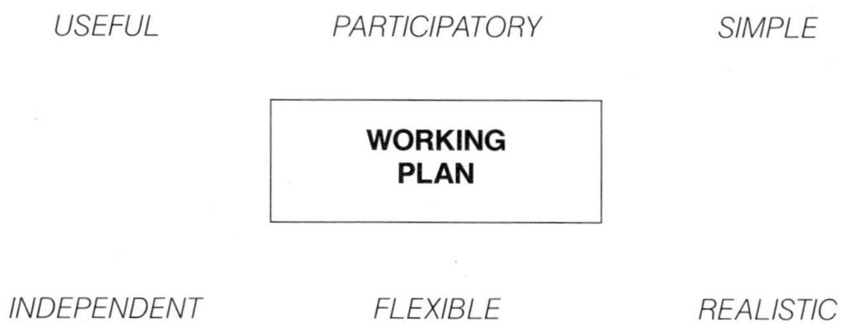

USEFUL *PARTICIPATORY* *SIMPLE*

WORKING PLAN

INDEPENDENT *FLEXIBLE* *REALISTIC*

Simple: easy to understand	Not containing technical terms which cannot be understood by users. For example, there is no need to include inventory data as these are not needed by the user group to manage their forest.
Useful: meets the users' objectives	Only describing forest management operations which need to be carried out to meet the users' objectives. The management objectives should also be clearly defined.
Independent: written by the users	Since the forest is under the management of the user group, they, as forest managers, should prepare the plan – not the ranger or the forest guard.
Realistic: can be carried out by the users	Users should decide whether they can really carry out an operation before including it in the working plan. Extensive operations should be carried out gradually in a small area each year rather than throughout the whole forest area at one time.
Flexible: can be altered if necessary	Since it is a working document, users will learn from experience what can and cannot be achieved and can change the plan according to these experiences. Major changes may need approval from the DFO.
Participatory: all users involved	This is difficult to achieve in practice. If working with the committee, ensure that all interest groups are represented and that the completed plan is approved at a whole group assembly.

Source: Branney and Dev, 1994a

with the ranger acting as facilitator. Ideally, the user group will have already chosen a representative committee and in the constitution described one of its functions as being the preparation of a forest working plan (see Box 6.7). In this case the ranger can work closely with the committee rather than the entire user group. From a practical viewpoint, this is important, especially if the user group is very large.

The first stage in the process is the preparation of a detailed sketch map of the community forest, recording on it all features of relevance to the user group such as houses and settlements; trails; streams and rivers; temples; and other physical features. This is done through a participatory mapping PRA exercise, preferably taking place in the forest itself, or at a point from which the whole forest can be viewed. Survey maps, if they exist, can sometimes be used to provide the outline or external forest boundary on to which other features are sketched by the users. The forest area itself is divided into blocks (**bhag**) based on areas of similar type and condition, with easily recognisable boundaries. These blocks are sketched on to the map and named or numbered by the user group. An example of a sketch map is shown in Figure 6.3. Blocks can be of variable size, but should be identifiable as part of the forest requiring a distinct set of management prescriptions.

Once the forest area has been divided into blocks, each block is visited in turn by the user group committee and the ranger. Providing technical advice if required, the ranger encourages the committee to describe and discuss each block in turn, and assists in recording any decisions made. For each block, the working plan describes the following:

Block description	Name; boundaries; forest type; condition; main species; age; area estimate, etc.	Described in terms understandable by all users, e.g. qualitative description of forest condition; local names; local measurement units.
User group's objectives	For example: to improve forest condition; to harvest fuelwood; to generate income, etc.	Users should set their own realistic objectives for each block.
Forestry operations	A description of the operations needed to be carried out to meet the objectives, e.g. coppicing; enrichment planting; timber harvesting, etc.	Users select operations from a series of options suggested by the ranger.
Type of products	Based on the operations described, the type, quantity and time of availability should be described.	Quantities may be difficult to specify until the user group has more practical experience.

Figure 6.3
Sketch map of Mainakhop Giddhekhop Community Forest, Dhankuta District

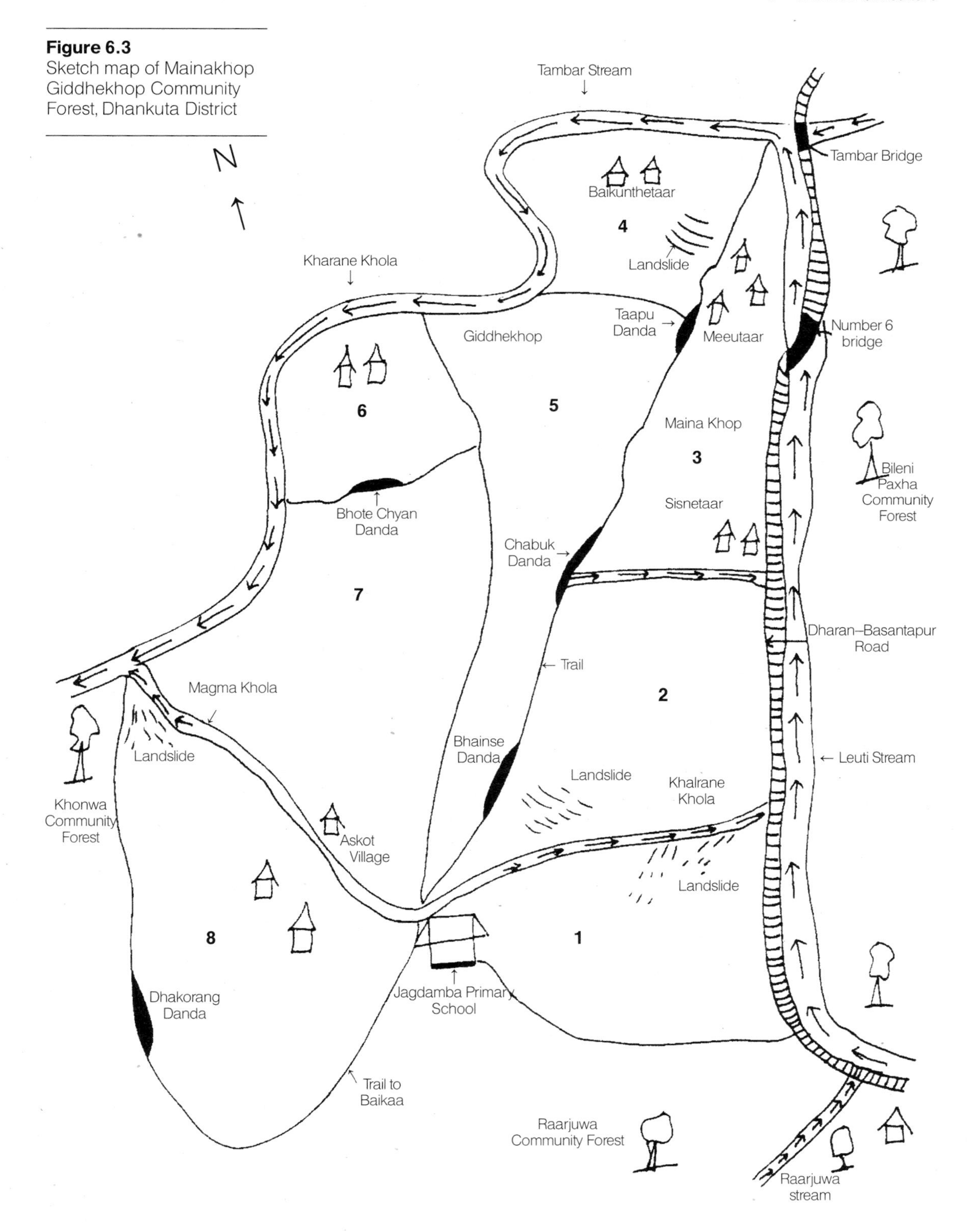

Box 6.8 Translated Section of the Working Plan for Mainakhop Giddhekhop Community Forest User Group, Mahabharat Ward 6, Askot, Dhankuta District

Introduction to the forest

In the past, this forest contained very large trees where vultures (**giddha**) and myna birds (**maina**) made their nests. Thus, the forest was called Mainakhop Giddhekhop. It is situated to the west of the Dharan-Basantapur road. This forest occupies an area of about 125 ha and contains plants and trees of **karam**, **sakhuwa**, **khayer**, **badkule**, **hade**, **saj**, **chiuri**, **satisal** and **jamun**, and wildlife such as porcupine, monkey and wildcock. The forest contains plants and trees ranging from small naturally regenerated saplings to large mature trees usable for timber. The forest as a whole can be taken as a mixed-age woodland.

Overall objectives of forest management

- to make the forest increasingly productive to cope with the increasing population
- to control the uncontrolled use of the forest
- to carry out forestry operations

Block number 1

This block borders on Jagadamba Primary School, Askot in the west; Dharan to Dhankuta motor road and Leuti Khola in the east; Kharane Khola in the north; and Khairala forest in the south. Extending to an area of about 60 ropanis (3 ha), this block mainly contains many pole-size trees of **sakhuwa**. It also has useful species such as **hallude**, **karam**, **khayer**, **satisal** and **sindhure** in mixed form. Amongst the trees that are scattered about, there are shrubs and bushes of **dhangero**, **asare** and **tite jhar** and climbers of **bhorla**, **debre,** etc. Although this is mainly a **sakhuwa** forest, most of the larger trees are **hade** and there is plenty of natural regeneration of **sakhuwa**, **khayer** and **karam**. Along the Kharane Khola there is some possibility of landslides and erosion, hence protective measures are needed. The block has both steep and moderate slopes, it faces NE, and it contains dark stony soil.

Because of the possibility of landslides in this forest, emphasis will be placed on conservation work, with the promotion of natural regeneration to prevent landslides. In this way, the capacity of the forest will be gradually improved and any forest products will be distributed and used properly.

In conjunction with the forestry operations, bushes and climbers will be cleared away, and bamboo and amliso grass will be planted to provide an income to villagers where landslides have stabilised. Pole-sized trees will be pruned. Branches and dry trees can be cut to yield fuelwood, which will be distributed fairly for the use of group members. The following working schedule has been devised:

• Weeding	June–July 1995	by all user group members
• Lopping and pruning	Dec–Feb 1995	by all user group members
• Bamboo and amliso planting	July–Aug 1995	by all user group members

Notes: Karam (*Adina cordifolia*); sakhuwa (*Shorea robusta*); khayer (*Acacia catechu*); hade (*Lagerstroemia parviflora*); saj (*Terminalia alata*); chiuri (*Aesandra butyraceae*); satisal (*Dalbergia latifolia*); jamun (*Syzygium cuminii*); dhangero (*Woodfordia fruticosa*); sindhure (*Mallotus philippinensis*); hallude (*Lannea grandis*); bhorla (*Bauhinia vahlii*); amliso (*Thysanolaena maxima*); other species not identified.

The result is a block-by-block description of the forest and of the management which the user group intends to undertake to meet its objectives. An example is shown in Box 6.8. This, along with the sketch map, is then presented to a users' assembly for agreement (with any amendments). A copy of the plan is also taken to the DFO for final approval. This is particularly important, since users may be unwilling to carry out more active forest management operations, such as harvesting of green wood, unless they are confident that these have the sanction of the DFO.

Community forest management workshop

The participatory method of preparing a working plan with a forest user group requires a new range of skills and attitudes on the part of Forest Department field staff. For several years now reorientation or start-up workshops have been successfully used in Nepal to enable field staff to adopt appropriate participatory approaches to community forestry (Gronow and Shrestha, 1990), with an emphasis on user group formation. NUKCFP has now developed a community forest management workshop to assist rangers and forest guards in recognising and implementing their role in supporting user groups to implement more effective forest management (Dev, 1994).

The workshop aims to develop the following main skills:

- Implementation of the participatory process of working plan preparation
- Use of PRA and local knowledge for describing the forest and forest management operations, based on visual observation rather than inventory and survey
- Application of appropriate management operations for natural hill forest, rather than an emphasis on plantations
- Incorporation of a broad range of forest management operations into working plans, including: agroforestry (e.g. grass, bamboo and cardamom planting); soil conservation (e.g. grass planting, checkdams); NTFP management (e.g. medicinal plants, resin tapping, lokta (*Daphne bholua*) harvesting)
- Methods for estimating product yields from community forests, based on self-assessment by user groups and harvesting demonstrations

The workshop is conducted in a participatory environment rather than through formal training, recognising that field staff often have considerable experience of working with user groups and the two sides can learn much from each other. Participants are encouraged to discuss specific situations involving community forest manage-

Figure 6.4 Pine Forest Management Chart

FOREST AGE	+ CANOPY DENSITY	+ REGENERATION	+ FOREST CONDITION	+ POSSIBLE MANAGEMENT OPTIONS
MATURE	Dense	Abundant	1. Very good	Selective felling; thinning
		Scattered/few	2. Very good	Selective felling; promoting natural regeneration; weeding/cleaning
	Open	Abundant	3. Good	Selective felling; pruning
		Scattered/few	4. Average	Promoting natural regeneration; pruning; weeding/cleaning
POLE STAGE	Dense		5. Good	Thinning/pruning
	Open		6. Poor	Pruning; enrichment planting
ANY	Very open	None	7. Poor	Promoting natural regeneration; enrichment planting; weeding/cleaning
			All conditions	Agroforestry; soil erosion control; fire control; grazing control; NTFP harvesting (resin tapping)

ment in their own working areas, and emphasis is placed on field visits to various community forests for practical sessions.

In view of the shortage of relevant written material suitable for field staff working in remote areas with forest user groups, guidelines have been produced by NUKCFP (Branney, 1994; Branney and Dev, 1994a). These describe a number of easily followed steps which enable rangers and forest guards to provide user groups with more appropriate management advice. For example, based on the users' own assessment of forest type and condition, and their own management objectives, certain management options can be provided (see Figure 6.4). These guidelines are now available in Nepali, and have been distributed to field staff in the project districts.

Forest management demonstration plots

The existence of a working plan prepared by a user group for its own forest does not automatically guarantee that the provisions of the plan will be implemented. In the case of management operations which require the harvesting of forest products, especially green wood, further support is usually required from field staff. As a means of providing this, NUKCFP has initiated a system for establishing forest management demonstrations (see Box 6.9). Sites are selected for these on the basis of a number of criteria, including the desirability of demonstrating active forest management in a range of geographical locations and covering a range of forest types and conditions. It was also decided that demonstration sites should be reasonably accessible if they were to have further value during user groups' networking workshops, and study tours, and community forest management workshops.

Forest management demonstrations are a type of action research in community forest management. The operation to be carried out is determined by the user group; in effect, this means that it has already been described in the working plan. The actual harvesting is then carried out by users, and the product yields are also measured by them. The user group can subsequently decide whether it wishes to repeat the operation, or modify it in some way to suit its own particular requirements.

Initially, the role of the ranger is to discuss with the user group the operations it has described in its working plan, and to select a relatively small area (usually 0.2–1.0 ha) within the forest block allocated. Users are then encouraged to carry out the operation within this fixed area. If some form of harvesting is required (coppicing, thinning, pruning, etc.) the users themselves (or more usually the user group committee) will choose the actual trees to be cut, with the ranger being on site to provide detailed advice.

Users are encouraged to record the amount of produce harvested from the area; most user groups already have some system for doing this according to their constitution. This information is of

Box 6.9 The Advantages of Using Demonstration Plots

As a focus for encouraging user groups to implement more effective forest management, demonstration plots have proved to have a number of benefits, including:

- Users' appreciation of the quantities of forest products they can obtain from forest of a given type and condition, leading towards more sustainable forest management.
- Increased confidence of user groups in carrying out harvesting operations with the sanction of Forest Department field staff.
- Improved practical experience of FD field staff as a result of being closely involved with forest management, particularly harvesting operations. This enables them to provide more effective support to other user groups who wish to manage their forests in a similar way.
- Demonstrations have been used as a focus for field visits organised for user group members during networking workshops, as a more effective means of sharing experiences and encouraging active forest management than more formal training of individual user groups.
- Accumulation of information on user group forest management (including yields of forest products) which in the longer term will assist the DFO in monitoring user group forest management activity.
- Establishment of a series of sites which can be used as 'tools' for further forest management workshops or training exercises.

value to them in undertaking future harvesting operations of a similar kind. Quantities are measured in easily understood units (e.g. backloads of fuelwood). Additional quantitative information is collected by the project and field staff who carry out a simple survey before and after harvesting. This enables data from different sites to be compared, and adds to the general body of available knowledge concerning user group forest management. Note, however, that no attempt is made to produce statistically valid data, or to replicate treatments, since this is beyond the scope of this type of action research. More recently, similar demonstration plots have been established covering other types of community forest management activity including grass planting with amliso (*Thysanolaena maxima* or broom grass) and the establishment and management of other NTFPs.

Characteristics of user groups as forest managers

The participatory forest management process, with its associated components as described, has demonstrated that user groups can be assisted in moving from a largely protectionist role to become effective forest managers. The process is not, however, without its problems – some of them are mentioned here. One of the most interesting aspects of helping user groups to develop as forest managers is that the type of management they wish to carry out in their forest is often somewhat different from that which would normally be carried out by trained foresters. However, since the entire participatory forest management process is at an early stage, with user groups still developing their skills and experience, much remains to be learned, regarding both the abilities of user groups to optimise the benefits they obtain from community forests, and the implications of this in terms of changes in forest condition and structure.

Given the huge range of variation between forest user groups, both in terms of household numbers and ethnic groups represented, and in the size, condition and type of forest handed over to them (Table 6.2), it is not surprising that there is a corresponding variation in the level of interest, skill, and effectiveness with which these groups manage their forests. Some user groups appear to be able to adapt more readily than others to becoming effective forest managers, and respond well to the post-formation support processes described. During a review workshop involving DFO field staff and NUKCFP staff, the criteria which seemed to have a major influence on this were identified and discussed with a view to prioritising the support being provided to user groups within each Range Post area (see Table 6.3). It appears that those user groups with a limited area of the more useful forest types (usually broad-leaved forests which can be managed as coppice) are those which most readily become effective forest managers through preparing and implementing a working plan using a participatory process.

Success of user group management is dependent on the forest type and quality and its ability to provide multiple benefits over a period of time.

Community forests managed under coppice or coppice-with-standards systems lend themselves readily to the production of a wide range of easily harvested products which can be utilised by individual households. High forest requires more organised harvesting operations, and more sophisticated tools and equipment, and produces material which could be marketed outside the user group (particularly timber). This means that high forest management is more difficult to implement, both technically and institutionally, by a relatively inexperienced user group. In addition, forests at different altitudes will also need to have tailored silvicultures developed for them. For example, work carried out by Jackson *et al.* (1993) in the high altitude forest of Kabhre Palanchok indicates that there is potential for the development of user group forestry. However, since users are dispersed and are generally involved in the collection of high value products, the types of systems developed for proximate users collecting firewood and fodder may not be appropriate.

Different management and organisational structures are required for forests where users are dispersed.

Newly formed user groups may take some time to acquire the confidence to implement more active management operations. However, it appears that the actual process of preparing the working plan, plus the increased forest product yields generated, can itself lead to a strengthening of the user group as a community-based institution representing the genuine interests of all the users.

In other situations, where user groups control extensive areas of forest with as yet plentiful supplies of dead wood which can easily be collected, not surprisingly it has proved difficult (and unnecessary) to encourage harvesting of green wood or even to control grazing.

Table 6.2 Forest user group characteristics in the Koshi Hills

	District				
	Bhojpur	Dhankuta	Sankhuwasabha	Terhathum	Mean/total
Mean community forest area (ha)	40.6	30.4	54.9	47.8	41.4
Range (ha)	0.5–400	1–150	1.3–400	1.5–750	0.5–750
Mean number of households per FUG	87	98	78	82	88
Range	11–351	15–392	16–254	20–219	11–392

Source: Gayfer (1994)

Table 6.3 Criteria affecting the effectiveness and interest of FUGs as forest managers

Most Effective	Least Effective
Remote (no access to markets)	Near bazaar or motorable road (access to markets)
Near range post (frequent contact with field staff)	Distant from range post (little contact with field staff)
Small forest area (easy to manage)	Large forest (harder to manage)
Many users per unit forest area (high resource pressure)	Few users per unit forest area (low resource pressure)
Homogeneous user group (fewer disputes)	Heterogeneous user group (more disputes amongst members)
Shorea robusta and *Schima-Castanopsis* forest (can be managed under coppice for a range of forest products)	*Pinus roxburghii* forest (high forest the only real management option)
User group established for some time (more confident)	Newly formed user group (less confident)

The following criteria seemed to have little or no effect:
Forest condition (good/average/poor)
Forest origin (natural stand/plantation)

Forest product yields and types of products

From demonstration plots, and from information collected during user group networking workshops, it is now becoming possible to accumulate information on yields of forest products resulting from harvesting operations carried out by user groups (Branney and Dev 1994b). This information can be used in two ways: firstly by the users themselves, to modify and amend their forest management prescriptions, and secondly, by Forest Department field staff for supporting and monitoring user group activity.

Implementation of a working plan by a forest user group can result in a significant increase in the group's output of forest products. The yields actually obtainable lie somewhere between those being obtained under the 'no-management' or protective regime and the maximum biomass output which is theoretically possible under controlled research conditions. Although user groups may not be trying to maximise their total output in terms of quantity of products, they can be encouraged to harvest in a more systematic way, thus providing more benefits than if the forest was simply protected from all green wood cutting. As a result of harvesting within a demonstration plot, and seeing the immediate and longer-term effects on the ground, a number of user groups have modified their working plans to allow themselves to harvest more fuelwood or other forest products.

Whereas foresters are traditionally concerned with the production of a single, or limited range of, forest products, forest user groups are more concerned to manage their forests to maintain the output of a much broader range of products including fuel, fodder, poles, timber, leaf litter, food, agricultural and household implements, grazing and medicinal and religious plants. Bearing in mind that within each of these categories there are a number of plant species

Measuring the yield from a pine plantation, Nepal

Oak regeneration maintained at waist to shoulder height to facilitate fodder cutting, Nepal

which have different qualities and which are required at different times of the year, it is clear that community forest management can become very complex, and is likely to be somewhat different from management for a more limited range of products.

An interesting example of this arose recently when it was found that in the higher altitude oak (*Quercus* spp) forests, although *Quercus semecarpifolia* is a major fodder species, at certain times of the year the parasites, epiphytes and climbers which festoon this type of forest are even more important for fodder when the nutritional value of oak is low. These plants, which would be considered non-productive or even damaging by foresters, are an essential part of the complex subsistence agriculture-forestry interface at this particular altitude, and user groups will aim to manage them in the same way as they manage their forest to produce fuelwood or poles.

In another situation, user groups following a coppice management system within a *Schima-Castanopsis* forest did not cut as many of the trees and shrubs within a small coupe (coppice area) as would be desirable in order to maximise biomass production. Certain plants were retained for future timber purposes (*Schima wallichii*), for pole production (*Castanopsis* sp.), for medicinal use (*Mahonia nepalensis*), or for bark production for Nepali paper making (*Daphne* sp.). This leads to a less systematically managed coppice system, but nonetheless is a system which is better able to supply the mixed product requirements of the user group.

Technical assistance and information requirements

Handing over forests to user groups to be managed in accordance with the complex requirements for multiple forest products does not imply that Forest Department field staff no longer need to be concerned with forest management themselves. The support of field staff in facilitating this process is critical to its success. In addition, although user groups may have a wealth of indigenous knowledge concerning forest product uses, they do not necessarily have the answers to all the problems they encounter, and requests to the District Forest Office for assistance increase rather than decrease as these groups become more involved in active management.

User groups are now starting to seek advice on the quantities of forest products which can be harvested from their forest. This is often a recognition that improved forest management is needed to satisfy the demand of the group members. By collating information from harvesting demonstrations, it is becoming possible to increase the body of knowledge available to field staff and user groups.

In contrast, DFO field staff are required to monitor user group activity to ensure that community forests are being sustainably managed, and that the provisions of forest working plans are being followed. Again, by collating and adapting information from actual situations where user groups are managing forests, it is becoming possible to draw up some simple rules of thumb to assist Forest Department staff to do this.[3] Clearly, the gaps in present knowledge concerning user group forest management are still large. Through a collaborative process involving both user groups and field staff, and assisted by NUKCFP, actual forest management experiences are being interpreted and translated into a format which can feed back into the participatory processes of management workshops and working plan preparation, to stimulate and support more effective forest management.

6.4 Discussion

As this review of some of the emerging issues and experiences underlines, there is much still to be learned about appropriate silvicultural techniques in diverse and complex ecological and social systems. Perhaps one of the outstanding questions concerns the impact of different silvicultural options on the livelihoods of forest-dependent groups, and the development of mechanisms to monitor this impact and to ensure corrective action if necessary.

As user group management plans are developed the anomalies between them and the traditional Forest Department working plans are emerging. In a sense this signifies the appearance of the real contradictions between different forms of working: where bottom-up plans based on meeting multiple objectives conflict with plans derived from a top-down identification of often single objective management approaches. The resolution of this fundamental contradiction remains an extremely urgent issue for all involved in the development of a meaningful approach to local forest management.

Notes

1. Material in this section draws on Campbell (1995) and Rathore and Campbell (1995).
2. For a comprehensive discussion of the experience to date with new approaches to participatory resource assessment see Carter (1996).
3. For further information on some of the innovatory resource assessment techniques developed by NUKCFP and the Nepal Australia Community Forestry Project, see Carter, 1996.

7 The New Institution

> Among the greatest futilities in development efforts are the unsuccessful attempts to bring about improvements in institutional efficiency through training and improved management techniques that run counter to the incentives engendered by the policy framework (Ostrom *et al.*, 1988)

7.1 Introduction

Chapter 4 considered the existing and new local organisational partnerships that have been developed to manage forests. In this chapter the other side of the partnership – the Forest Department is considered in terms of the impact on these agencies of the devolution of control to a local level. Further consideration is also given to an assessment of the mechanisms developed for ensuring bottom-up planning and linking decision-taking with the new bureaucratic structures. The role of external donors, their project advisers and NGOs is also considered as a mechanism to facilitate institutional change.

7.2 Institutional Restructuring: The Institutional Continuum

Challenging the formal institution: the role of government agencies

Government Forest Departments, as large bureaucracies, have organisational characteristics that both support and run counter to the successful institutionalisation of decentralised forest management (Kant *et al.*, 1991; Rastogi, 1995; Roy, n.d.). (Table 7.1 highlights some of these constraints.) On the one hand, they can provide a long-term base of resources and decision-making capacity accountable to the public interest. On the other, experience has shown that programmes administered by bureaucracies tend to become more rigid and top-down, especially as they expand. Decentralisation

Table 7.1 A problem list of forestry-related institutional constraints

General to all bureaucracies	Specific to forestry institutions
Overcentralisation of bureaucracy undertaking development	Historical trend to accumulate more power and control over forest resources as doers through territorial approaches rather than working as facilitators in partnership with local users
Lack of preconditions for participatory approaches	Poor training of staff to work with people; all forest agency staff are foresters rather than a range of specialists working in interdisciplinary teams on socioeconomic and technical issues
Lack of specialisation in career streams	Lack of specialisation in career streams for innovative areas, such as community forestry, silvo-pasture development, joint forest management, agroforestry, enterprise development, extension
Lack of flexibility in staffing depending on responsibilities	Orientation to timber and major commercial products rather than on multiple uses and multiple users including the wide variety of products and processing technologies and scales
Lack of criteria and flexibility to transfer responsibility to local people, private sector, NGOs	Poor attention to local control and management capabilities for common property resource utilisation; lack of attention to market and income generation potential; lack of linkages to local communities and user groups, private sector and NGOs
Narrow planning within sector, when many issues such as land use are cross-sectoral and solutions require cross-sectoral action	Forests are more generally perceived as a land reserve or as a residual use of land, thus leading to a policy and legal framework that is a disincentive for positive forestry development; lack of dialogue between forestry and agriculture and other sectors; inability to deal with land tenure and use issues; absence of incentives for extracting economic rents
Agencies responsible for cross-cutting areas are not properly co-ordinated or staffed for issues and problem analysis	Staff performance incentives based on physical targets rather than demand-driven accomplishments or sustainable forestry development
Local-level administrations not geared to undertake development responsibilities	Planning system too complicated for participatory involvement of non-technical foresters, more flexible time-frame for forestry operations
Inefficient and untimely flow of funds for programmes	Longer gestation for forestry programmes creates funding problems. Time frame for meeting targets too inflexible for participatory tasks

Adapted from: Gregersen *et al.*, 1993

policies are, however, leading to a slow internal restructuring of formal institutions where lower-level staff are being given increasing responsibilities for large elements of management. However, as with local groups where devolution of power has been partial, so it is within the government institutions. Individual innovation is unlikely to be rewarded, where incentive structures are predicated on observing the hierarchical norms of behaviour (Bajaj and Sharma, 1995).

Innovation is a prerequisite for an organisation that is going to be able to respond to a dynamic environment, where the local-state interface has acquired a demanding voice. Innovation requires individuals to take risks, to learn from experience, and to be able to admit failure. The structure of bureaucracies rewards risk-averse behaviour that conforms to norms accepted within the institution. This leads to the situation facing most Forest Departments today, where the political and social context increasingly requires a responsive, accountable, innovatory learning organisation, but instead is left with the opposite of all these desired characteristics. How, then, do public institutions move from one end of the cultural spectrum to the other? What are the incentives to support this type of change? Does participatory forestry provide an appropriate entry point from which to begin the transition to a new institutional context? The following sections look at the role of the external agency in facilitating the change from old structures to new, and consider both the open and hidden relationships within organisations that may provide constraints to change.

In Table 7.2 a series of issues facing the forest sector is highlighted, together with a comparison between the existing or 'traditional' approach to management and the new desired responsive organisation.

In many cases, by the end of the project the organisation with which it has been working has ostensibly started the transition from old to new. All too frequently, however, shortly after the withdrawal of project funding and staff, organisations return to their old roles. The aim of new forestry projects is to make overt the focus on organisational change and to support this process directly rather than as an *ad hoc* additional benefit that may or may not occur. Thus, new projects focus on means rather than ends; on identifying and involving clients or stakeholders as partners rather than beneficiaries; on setting objectives and finding solutions (problem solving) rather than implementing blueprints; on learning from experience rather than assumptions.

The role of the external agency

Following a typology described by the FAO in 1980 (Box 7.1), the form donor projects may adopt to facilitate change is described in this box – 'institutional development model'.

In India the external facilitator is not usually an expatriate, but is

Table 7.2 Participatory forestry as a lever for organisational change: from old to new

Issue	Traditional approach	New approach
Client/interest group focus	Inadequate Based on assumptions	Essence of approach
Understanding about client/site requirements	Standard assumptions/packages Higher-level management 'knows best'	Too much variation for standard assumptions to be applicable The person nearest to the client/site knows best – i.e. the lowest-level field staff
Gender	Little participation	Women seen as one of major client/interest groups and project partners
Standard packages (models)	Common to all situations Based on expert opinion	No standard packages Measures are site-specific or client-specific Guidance and training are provided to staff and responsibility and authority devolved to them to make appropriate decisions
Decentralisation of power	Most decisions taken at higher levels of organisation	Devolution of power to field level
Responsibility sharing	Department only	Department and public
Human resource development	Based on position in hierarchy Standard skills assumed Leads to lack of motivation	Needs-based assessment Aim is to improve quality of all aspects of work
Attitudinal change	Very rigid (militaristic) Staff afraid to take initiative Questioning/innovation discouraged	Responsive Senior management commitment and support required
Control and monitoring	Focus on control *'Can't Do'* mentality	Focus on planning, innovation, analysis and change in practice *'Can Do'* mentality
Shared understanding of goals, objectives and risks	Not needed Issue instructions and follow orders according to long defined manuals and models 'Need to know' basis only	Essential Joint development of objectives, expression of risks Team approach Respect for individual skills and knowledge

Source: Shields, 1994

an Indian national often working in an NGO or university research programme. There are obviously exceptions to this but the relative importance of non-nationals in the India programme is significantly less than in Nepal where donor interest has sometimes been likened to the establishment of separate fiefdoms across the regions of the country. However, as can be seen from the discussion in Chapters 4

Box 7.1 Project Organisation Form and Function

Supermanagement model: in which the forestry administration manages the project for the village (social forestry and earlier phases of community forestry).

Support service model: in which a community organisation manages the project and the forestry administration contributes technical knowledge and supplies (some examples of joint forest management and community forestry).

Partnership model I: in which the forestry administration and the community form a joint organisation to manage the project (JFM (in rhetoric) and some examples of community forestry).

Partnership model II: in which an outside organisation (e.g. a rural development agency or a voluntary organisation) acts as an intermediary in facilitating or implementing the joint management of the project (examples in India and Nepal).

Institutional development model: an advanced form of local institutional management (where the focus is on the institutional development of government and local organisations).

Adapted from: FAO (1980)

and 5, the role of donor projects and volunteers has been influential in helping to develop a vibrant and responsive form of participatory forestry in Nepal.

The volunteer programme in Nepal has been an important source of outside support to Forest Department work in community forestry (including volunteer programmes supported by Britain, Denmark, Germany, Japan, the Netherlands and the USA). Volunteers work within the Community and Private Forestry Division and are part of the Community and Private Forestry Programme. Volunteers, as opposed to project staff, have an unusual and often difficult position in that they are working alone, and usually with no formal authority over staff or budgets, at the District Forest Office in support of the DFO's programme. When the DFO is interested, then good relations develop. However, in some cases, volunteers continue a programme of work with local people irrespective of the interest or support of the DFOs (Box 7.2).

Roychowdhury (1994) analyses the impact of external agencies as agents of reform. In this instance the World Bank has been able with large sectoral funding to State to use financial leverage to enforce structural change in several State Forest Departments in India. However, is such external enforcement necessarily going to lead to the type of substantive change required by the sector? As Madhu Sarin, a leading analyst of forest sector changes, is quoted as saying (see Roychowdhury, 1994):

> Some of the changes suggested by the Bank might seem all right – and even desirous in principle – but the implication of the Bank pushing for these is worrying. The initiative should have come from the government here.

Box 7.2 What Does a Volunteer Do?

There are few clear-cut roles for volunteers since circumstances vary enormously. The major determining factors are the DFO's understanding and commitment to community forestry, his management abilities, the interest and aptitude of the rangers for extension work, the number of rangers, the location of the district and its infrastructure.

The main role for volunteers is to tailor their work approach to their districts and to focus on laying the foundations for community management of forest resources.

Roles of the Volunteer

- catalyst
- trainer
- facilitator
- implementer
- link to outside funding opportunities
- link to outside technical advice and resources
- innovator

Volunteers have encouraged user groups to set up their own self-financing training courses; for example, one user group in the Tarai organised a one-day pruning, nursery and tree planting course where participants were charged Rs20 for the day. In other cases volunteers have organised training courses on building improved cooking stoves and fire management, which are now run by members of user groups without inputs from either volunteers or the forest office.

The DFO's View

This statement from a District Forest Officer indicates his view of the importance of the role of volunteers:

The DFO in Darchula worked with a volunteer there for one and a half years. He has met other VSOs also and is positive about the contribution volunteers make to the community forestry programme. He feels that volunteers are able to devote time to working with local people, to inform them about the various forest management options and to help them to plan what they wish to do with the forest. Often DFOs and their staff are too busy with administration and public relations work and cannot spend sufficient time developing the community forestry programme. The DFO suggested that the volunteer has an important role as an information channel between local people and Forest Office and can also act as a community forestry adviser to the DFO.

Source: VSO volunteer reports

The fear that donor agendas and enthusiasms will not lead to the internalisation of these approaches is mirrored by Singh and Khare (1993), who suggest that the donor imperatives will lead to 'target and time-frame chasing only'.

Rationalisation of the administrative structure and functions at each level has been proposed by donors, in response to the new forms of forest management. In the social forestry era, the donors insisted on separate social forestry divisions; in some States there are now separate JFM units, in others JFM support teams, in yet others JFM responsibilities are performed by the territorial staff. Some donors have suggested that new participatory forms of forestry will allow Forest Departments to reduce staffing levels, since many of the duties undertaken by field staff will become the responsibility of villagers. As can be seen from Box 7.3, the new approaches bring many uncertainties, and it is a far from simple transition from the old-style bureaucracy to a new responsive agency geared to supporting the needs of client groups.

Forestry staff face a series of problems: 'low salaries, low incentives, low support, no job guarantee, no proper appraisal, and no clear-cut posting and transfer policy. These factors result in low commitment and dedication to community forestry, and to the profession and its institutions'.

(N.R. Baral, 1993)

As Saxena's acute observations highlight, there is an inherent problem in forcing the pace of institutional change beyond the ability and will of individuals within the organisation to change. Often the reasons for resistance are obvious but the project planners have been unable to listen, or to interpret official silences or passive acquiescence in its real light as one of unwillingness to be part of a project which appears to be impractical for a number of reasons. Griffin (1987), reflecting on his role as project director of the Nepal-Australia Forestry Project (now called the Nepal-Australia Community Forestry Project) notes:

> Most bureaucrats ... are keenly aware of the institutional framework within which they work: their promotion and survival depends upon this. Knowing the system well, it is no wonder that high senior officials in central offices and potential co-managers in the field look with scant enthusiasm on project proposals which take no account of practically important aspects ... Delays and prevarications may therefore sometimes be ultimately beneficial if they cause re-thinking or even cancellation of a proposal. If through donor enthusiasm an institutionally inappropriate project is implemented, it is likely to draw little more than perfunctory activity from recipient country staff. Expatriate staff will soon become frustrated or disillusioned by their inability to make the bureaucratic machinery work in ways for which it was not designed ... (T)he answer surely is to let the project and institution evolve together. If this is impossible, then the project should not commence.

Training and development of capacity

Despite the increasing realisation by donors that training alone will not lead to fundamental changes in organisations, much of the funding for participatory forestry programmes in both India and Nepal over the last decade has been to support the training of Forest Department staff. It has focused on providing staff with the additional skills necessary for the implementation of community forestry and JFM, particularly those concerned with improving interactions with local people (Malla *et al.*, 1989; Jackson, 1989; Gronow and Shrestha, 1990; Moench, 1990; Kant *et al.*, 1991).

The Ford Foundation, as part of the programme to support the development of capacity and to institutionalise and expand

Box 7.3 Forest Department Field Staff Misgivings about JFM Programme

- Lack of political commitment to creation of new organisations and empowering them.
- Lack of community interest in coming together and taking responsibility, as shown by the social forestry experience.
- In the Indian administrative system joint responsibility often results in diluted accountability. Only a single line of command can work in India. (This refers to the parallel structure established in some States where there are separate territorial and joint forest management staff both working in the same area).
- JFM may increase encroachments on forest lands.
- Indian villages are hierarchical, and collective management does not work well.
- Under international pressure, States are being pushed to increase their targets for JFM, whereas, with a new programme requiring high managerial input, progress should be slow and each State should draw up its own targets and plans.
- Many forest areas harbour 'undesirable elements of society', making it difficult to develop JFM approaches.
- Many NGOs lack the capacity to support the level of activity required under the programme.
- Long-term planning in forestry requires security of tenure and rights, whereas JFM may result in a dilution of existing rights.
- Political interference with staff postings does not allow a Forest Officer sufficient time on the job to develop the relationships necessary for implementation of effective JFM.
- The panchayat is too large an administrative unit to be entrusted with JFM organisation. Creating new organisations may cause conflict between the existing panchayat structure and the new forest management organisation.
- The new beneficiaries may be in conflict with the old beneficiaries under the Forest Settlements of the last century.
- There is the problem of free riders among the local people where many benefit from the activities of a few.
- JFM does not emphasise the important technical aspects of forest management, although technical issues will determine the nature of the outputs and thus the volume of benefits to be shared.

Source: N.C. Saxena (pers. comm.)

participatory forestry approaches, has provided support to a number of organisations to develop training programmes for Forest Department staff. In India, IBRAD, VIKSAT and MYRADA, amongst other NGOs, have been responsible for the development of in-service training courses (see Box 7.4). These have been highly successful in introducing new ideas and which to date have not been taught through the formal forestry training programmes. The NGO-run training courses use methods which emphasise the 'learning-by-doing' approach rather than the more usual 'chalk and talk' forms of teaching. This in itself has exposed Forest Department staff to new learning and information-sharing approaches. Similar experiences in Nepal have indicated the value of experiential learning approaches, and have been developed by several projects including the major community forestry training programme supported by Denmark which operates throughout the country (Gronow and Shrestha, 1990).

In India, dissemination of information about joint forest management, through a national-level NGO, the Society for the Promotion of Wastelands Development, to lower-level Forest Department staff has provided them with access to ideas and experiences which previously were only accessible to staff at more senior levels. This positive experience needs to be tempered by recognition of the effects of patron-client relationships found in most government organisations in South Asia (Fox, 1991). The introduction of certain forms of technologies and training may not be able to transcend these hierarchical relations, and in many cases lower-level staff who

Box 7.4 Development of Forest Department Capacity: The Role of NGOs

IBRAD has excelled in its work of training and re-orienting forest officers, evolving strategies to re-orient staff from the top-down training programmes, combining behavioural psychology with role playing, case study materials and field exercises which expose officers to the personal rewards of participatory planning and management. Building on this expertise, IBRAD has become a national expert in this area, sought after by many States and the national government to help set up training programmes for joint forest management.

VIKSAT has worked on training programmes for field-level forest staff and with communities. This has led to their increasing interest in community institutions building and to facilitating a growing informal 'association' of community JFM institutions. Convened as an experiment with 5 Village Tree Growers Co-operatives, with whom VIKSAT worked, the 'Parishad' now includes over 20 co-operatives which meet faithfully once a month. A small newsletter now serves this group, and exchanges with other village institutions involved in JFM in other parts of the State are beginning to occur. As an outgrowth of this activity, VIKSAT has undertaken research in JFM village institutions in different parts of the country and is evolving a theoretical framework for analysing their effectiveness. Both these grants illustrate the strength and dynamism necessary for a facilitating institution to evolve with the needs of the groups it is helping.

Source: Jeff Campbell, Ford Foundation pers. comm.

most need exposure to new skills are not provided with them, or if they are, the structure of the organisation remains inimical to the implementation of what they have learned.

An example of this problem is provided by the new culture of information management, heavily promoted by international organisations, which sits unhappily where access to information is privileged and cannot easily be relinquished to lower levels of the hierarchy. As Fox (1991) states: 'other problems arise when officials who possess information view it as a scarce resource to be exchanged for scarce commodities or influence of equal or greater value'. Thus, as calls for 'bottom-up planning' increase from lower-level forest officials, senior management responds by increasing control over information flows. The increased use of centralising technologies such as sophisticated Geographical Information Systems may ensure that those who are not computer-literate have even less access to decision-making within the bureaucracy. The development of capacity within an organisation is not just a straight transfer of skills and technologies but requires an environment in which staff are able to respond and are allowed to put into practice what they have learnt.

In addition to the problems of restructuring government organisations, there are, as yet, unquantified costs involved in developing participatory management of resources. These costs will increase according to the hierarchical complexity of the organisation and thus the number of tiers of decision-making, unless authority for decision-making is decentralised. Participation costs are related to three factors (Picciotto, 1992):

- the number of communication links between levels in the organisation
- the frequency of transactions carried by these links
- the average intensity of these transactions

In a centralised decision-taking organisation the costs of participation are very low. Decisions are taken by a small number of individuals and passed down the system to subordinate staff without question. However, this is an oversimplified view of very much more complex decision-taking structures. At a superficial level and in a very real sense, participatory decision-taking structures do necessitate large amounts of additional staff time and resources to ensure that the quality of participation is maintained. It is easy for such structures to degrade rapidly to a more centralising form of decision-taking where the number of actors and iterations is reduced.

The fragile relationship between field staff and local people can easily be destroyed where participation built on trust can be broken by staff who fail to honour their commitments. It is not just a question of the financial costs associated with participation but the human costs associated with changing attitudes towards local people and thus the level of respect accorded to them by field staff.

Almost all the forest protection committees had agreed with the range officer to hold their meetings on Tuesdays to ensure that he could attend. The meetings were held between 3 and 4 p.m., and provided a good opportunity for communication and review. Initially, this system worked well and the meetings were held regularly. Recently, however, although the group members gathered for the afternoon meetings according to the agreed schedule the range officer failed to appear. The villagers were not warned in advance and therefore have wasted their time and lost confidence in their relationship with the Forest Department.

(adapted from Roy *et al.*, n.d)

The development of participatory structures at village level thus has fundamental implications for the public sector in terms of structures, budgets and human relationships. The following sections consider some of the hidden relationships that condition interactions both within the Forest Department and between the department and local people.

'Much development effort is misdirected because of misdiagnosis. For example, many of the efforts directed toward institutional development focus on training and internal management systems. Yet this is likely to be ineffective if rent-seeking drives incentives in detrimental directions. We need to understand and undertake management improvements in the context of changing incentives'.

(Ostrom *et al.*, 1988)

7.3 The Hidden Institution

Although much innovative work has been carried out developing new interactive training approaches, it is clear that there is little fundamental change within the implementing organisations. Structures and functions remain the same, and many of the relationships that hinder the development of more responsive and flexible action are firmly fixed, regardless of the number of training courses undertaken or the style in which they are conducted.

Institutional norms and behaviour are governed by a series of rules and regulations that define the boundaries of acceptability. However, when considering different types of institutional arrangements, it is necessary to move beyond simple assessment of formally accepted rules to an assessment that takes cognisance of the hidden rules and incentives that actually provide the boundaries for individual behaviour.

Just as it is naive to assume that so-called people's organisations are necessarily a more equitable and desirable means through which to manage forests, so no government institution should be seen as homogeneous in its distribution of benefits and access to power. Each institution, whether formal or non-formal, is composed of individuals whose behaviour is governed by interactions that are both hidden and open. It is the hidden interactions, the most difficult for an outsider to comprehend and to incorporate within a programme of institutional change, that will determine outcomes (see Gilmour and Fisher (1991) for a discussion of these issues in Nepal). Rhetoric and planning can only address those overt structures that are amenable to discussion, for example policy and legislative frameworks, human resource development, and remuneration. Patronage systems, both within the service and with outsiders, are often the real determinants of institutional performance, condition all interactions and are the most difficult to tackle (see also Shah, 1991). As is seen in Box 7.5, questions concerning hidden relations have remained on the administrative agenda in India for well over a century.

The history of the development of the Forest Service in India is pertinent to the discussion of current performance. The division of the Forest Department into three separate services in which staff were derived from distinct sources has served to perpetuate inequalities and resentments between different levels. At the outset the following three services were identified (Buchy, in press):

- The Imperial Forest Service, staffed by the Conservator of Forests, District Forest Officers and Assistant Conservators of Forests who were responsible for controlling the implementation of forest policy. These staff were trained initially in France in the forestry school at Nancy, and subsequently senior officers were sent to Oxford to the Forestry School for their training. Later this training was replaced by that provided at the Imperial Forest School in Dehra Dun.

- The Executive Forest Service staffed by the Range Forest Officers who implemented forest policy. These staff were trained in the provincial schools based in Dehra Dun and Pune.

- The Protective Service, in which the remainder of the personnel were grouped, whose function was to protect the forests. Their training was restricted to learning from experience, with no other formal support provided.

From the earliest days these distinctions fostered resentment between the different levels of the forest service. The discrepancies in training, duties and benefits were also reflected in large wage differentials, with the lowest-level field staff, who were also the interface between local people and the government, on meagre salaries. This did nothing to encourage fair practice, and there are many records of field staff abusing their positions and extorting payments from local people. In addition to the wage disparities, the lower-level staff had no prospect of promotion from their rank to the executive service, which served to alienate them further. These distinctions continue today and lie at the centre of some of the major organisational problems.

Where power and status within a Forest Department are equated with control over a large area of forest territory, the 'soft' intangible control provided through participatory forestry provides little attraction for most professional foresters. This leads to a complex

Box 7.5 Hidden Relations and the Forest Department: a Century of Concern

The inferior subordinates of the Forest Department are perhaps as reliable as can be expected on the pay which we can afford to give; but their morality is no higher than that of the uneducated classes from which they are drawn; while the enormous areas over which they are scattered and the small number of the controlling staff render effective supervision most difficult.

It is not right, in order to protect the grass or the grazing dues on plots of waste scattered over the face of a cultivated district, to put it into the power of an underling to pound or threaten to pound cattle . . . .or to expose him (a subordinate servant) to the temptations which such a power holds out. Where the interests involved are sufficiently important, it may perhaps be necessary to accept the danger of extortion while minimising as far as possible the opportunities for it.

Source: IFS, 1894

operational environment for staff where incentives provided through the overt, formal structure are insufficient and unattractive to most of them. Just as was observed in the nineteenth century, levels of formal remuneration today are inadequate to meet the basic livelihood requirements of most government officials. Other ways of supplementing meagre incomes are inevitably found and in many cases these alternative methods become institutionalised. As participatory forestry increases the accountability and transparency of transactions, formerly hidden relationships are revealed, often to the cost of the individuals involved.

In addition to the lack of financial incentive to perform well provided through the formal governmental structure, externally funded projects, with their highly paid staff carrying out similar work to that of the government staff, are also causing tensions, as has been observed by several volunteers in Nepal working with district forest staff. This does however, raise questions of financial sustainability in the light of an already highly constrained public purse, and brings into question the donor policy of paying higher than government wage rates in order to secure good staff and/or enhanced performance.

Rates of pay for nursery workers:

- Himalayan Trust – Rs 2000 per month
- Care Nepal – Rs1200 per month
- HMGN – Rs500 per month

The problems associated with low rates of pay and poor conditions have been heightened by the move away from resource-creation projects (plantations) to institutional reform programmes. This change in project practice and funding has fundamental implications for project partners where highly centralised institutions are being asked to divest authority and control to lower levels both within the institution and to other organisations. Incentives (generally financial) provided through plantation programmes are no longer in place, and the demand for greater internal and external accountability is also putting pressure on individuals to change from hidden to open relationships (Box 7.6).

Box 7.6 Moonlighting and Daylighting

Clandestinely or openly, staff undertake economic activities. Often this is in agriculture and represents a 'strategy of necessity to remain professionals'. Salary supplements for work on aid agency or NGO projects are generally condoned by government and are common.

Extracting rents

(a) Subsidies, inputs, project funds are shared – a 'creaming off of surplus', often under quite fixed understandings and percentages for sharing. (b) Services are sold – foresters may be paid to move files, or provide documents, to carry out stock surveys, to designate taungya areas, to undertake log-scaling and grading.

Source: Chambers, 1993

Hidden budgets

In general, priority is given to activities for which there is a budget; if there is no money involved there is little interest. Hence, it is difficult to generate enthusiasm for village extension activities or devolved forest management where the budget is small and increasingly open to scrutiny.

In an example from India (Fernandes *et al.*, 1988), forest guards demand high bribes from local people, which they state are higher than they need be because they do not know how much money they will be forced to hand over to senior staff:

> The situation has deteriorated to such an extent that many interviewees only half jokingly suggest that the rates of each official should be fixed once and for all. Thus they would know exactly what they have to pay . . . Many officials said that if they knew exactly what the higher officials expected of them, they would not have to keep changing the amount they demanded (from lower level staff)

Although there are many such examples, it does not mean that the whole system is undermined by these hidden relationships. Indeed, project failure, as Chambers *et al.* (1989) state, is often due to mundane reasons caused by administrative weaknesses (e.g. budgets not released in time to allow planting to be carried out). However, staff, when discussing these issues, do state that the political pressure is increasing rapidly to operate under such relationships. Forest guards, as the main interface between local people and the government, face particularly difficult decisions as regards their position with respect to local people.

Indeed, the nature of 'corruption' should be considered in the context of the social relationships which construct it. Kondos (1987) and Gilmour and Fisher (1991) provide an interesting discussion of this issue, highlighting the relations based on caste and status (i.e. position, education, age and gender). Caste networks are often used by individuals to gain access to power and good jobs; status is used as a mechanism for domination and provides a hierarchical framework for interaction where everybody understands how they should respond to another (see Box 7.7). The use of **chakari**

Box 7.7 Patronage and 'Source-force'

The appointment of a person to a certain post in government frequently has less to do with his innate ability to perform in that position than with his **pahauch** (source-force). This refers to the contacts, the 'source' that a person has, and the 'force' that they can bring to bear to influence a decision. A person who has access to strong source-force has every chance of succeeding. . . A person with weak source-force has a much reduced chance of success, irrespective of actual ability.

Source: Gilmour and Fisher, 1991; Kondos, 1987

(usually interpreted as flattery) plays an important role in the development of social relationships and is used as a basis for reciprocal obligations. It is considered by all those who use it to be an essential social institution and is distinguished from acts of corruption (**brastachar**) (see Kondos, 1987 for a useful and full discussion of these systems).

These relationships are now coming into question as JFM focuses on the nature of the relationship and empowers different voices to question transactions between Forest Department staff and others. This leads forest guards to the unenviable position where, although they may often be drawn from the local population, they end up 'internalising the values of the system they represent and become enemies of the people', even though increasingly their work responsibilities expect them to be the 'friend of the people' (Fernandes *et al.*, 1988). However, as internal accountability and transparency within villages increase, the pressure to dismantle hidden relationships will also increase. This imperative for change will not necessarily coincide with the economic interests of the most politically powerful groups or indeed Forest Department staff who may have spent large amounts of money to secure a particularly lucrative position (see also Wade, 1988).

The target culture

Another facet of institutional perversity is the prevalence of the target culture, and assessment of performance on the basis of target achievement (Chambers *et al.* 1989). The problems engendered by this culture are common to both India and Nepal, and the constraints under which government officers work are similar, although more extreme for many officers in Nepal. District Forest Officers, like so many other government officials in Nepal, are often posted to districts far distant from their families. They work in a system driven by targets set at the beginning of each fiscal year. At the end of the fiscal year the DFO makes his annual report based on what has been achieved. He earns points on the basis of this report and these points influence his promotion prospects and his ability to be transferred to a more desirable district. In general, every DFO wishes to get closer to Kathmandu or to a roadhead; thus the more remote districts are considered to be punishment postings and any officer posted to them will be reluctant to take up post. To counterbalance this, more remote districts do gain additional points and there are remote area allowances to provide some compensation for the lack of services. However, such postings generally mean that officers with school-age children will not be able to bring their families with them since there will be little or no schooling of adequate standard, thus increasing the personal tensions associated with the job (see also Dahal, 1994 for a discussion of the impact of the transfer policy on officer commitment to support of user group forestry).

Apart from the pressure of the target culture, the organisational structure of the Forest Department in most States in India is not conducive to effective working with local people. Most departments are structured around securing the forest resource from local people rather than facilitating joint management. In order to reduce some of the pressures under which field staff operate, some senior Forest Department staff have recommended a complete restructuring of the department to ensure that there are more staff operating at the village interface level (Palit, 1993b). These recommendations have been supported by donor organisations which have also tried to facilitate the development of new inter-organisational linkages to provide access to skills and experiences not necessarily available within Forest Departments.

7.4 Institutional Experimentation

Experience from donors[1]

Experience from donor organisations is useful in helping to identify potentially successful new institutional arrangements. A recent review of the Ford Foundation community forestry programme highlights a series of mechanisms that have been successful in changing the working practices of Forest Departments in South Asia.

The last decade of Foundation-supported activities is marked by the development of new institutional partnerships. Forest bureaucracies' activities have become more collaborative through the development of stronger working relationships with other sectors of society, especially NGOs and universities. In these new relationships, each sector has drawn upon the diverse talents and experiences of the others to enhance its own work. For example, NGOs have been supported to organise communities, advocate policy change on behalf of forest villagers and provide technical assistance, legal and marketing services to villagers. Universities and research institutions have been supported to conduct training, develop new action research methodologies, analyse projects and provide documentation of project process. Direct support to government has been essential to create a sense of ownership and learning, and to enable agencies to develop new programmes and policies. Working together, often on the same projects at the same sites, has facilitated co-ordination and helped the members of these different sectors to develop mutual respect and understanding.

The 'three-legged stool'

Development of these multi-institutional linkages has been likened to constructing a 'three-legged stool'. All three legs of the stool – representing community-level action, policy formulation, and re-

Figure 7.1
Three-legged stool
illustration: Mark Agnew

search/training – are needed to achieve meaningful social change. The 'seat', or means of balancing and linking the three legs, is necessary as well. The collaborative relationships among the forestry department, other government agencies, NGOs, and universities or research institutions have enabled this integration to occur. It should be noted that any one agency or organisation may assume multiple roles in contributing to the three lines of work (as is seen in Figure 7.1). For example, both forestry departments and NGOs have played essential roles in organising communities and conducting research; universities have assisted in implementing programmes and contributing to policy recommendations. The organisations' effectiveness in performing these multiple roles has been enhanced as they collaborate with a broader array of institutions.

The role of NGOs: facilitators, advocates and implementers

If the demand for services driven by local organisations cannot be sustained by the government sector, can the non-governmental sector step into its place? (see Table 7.3) This debate has taken centre stage in the agriculture sector where there have been many examples of successful partnerships between NGOs and local people.[2] Joint Forest Management makes explicit reference to the need for partnerships to be developed between local people, Forest Departments and NGOs with NGOs (such as VSO) adopting a variety of roles depending on the particular need.

Similarly in forestry there are several successful examples of collaborative partnerships between government, non-governmental and local organisations. As Figure 7.2 highlights and Table 7.3 describes, a variety of relationships are entered into by GOs and NGOs which are dependent on the particular phase of development of a new idea (see Chapter 8) and the platform for action considered to be most appropriate by NGOs. It may also be the case that NGOs themselves change their role from one of advocacy to partnership as ideas become adopted by the state and the need for advocacy is removed. Other NGOs may wish to retain their independence from the state and retain a critical outsider view of the impact of interventions on different groups. A dynamic NGO-GO sector has elements of all these relationships.

For the last two decades NGOs in India have provided the voice of dissent and environmental conscience, placing the environment high on governmental agendas. The Chipko movement, the influential eco-feminist work of Vandana Shiva, in conjunction with the work of the Centre for Science and Environment and many others, has brought an enormous change in government rhetoric, and latterly in policy, as was discussed in Chapter 2.

There is another distinct group of NGOs that have had an even

Figure 7.2 Inter-institutional relationships in JFM

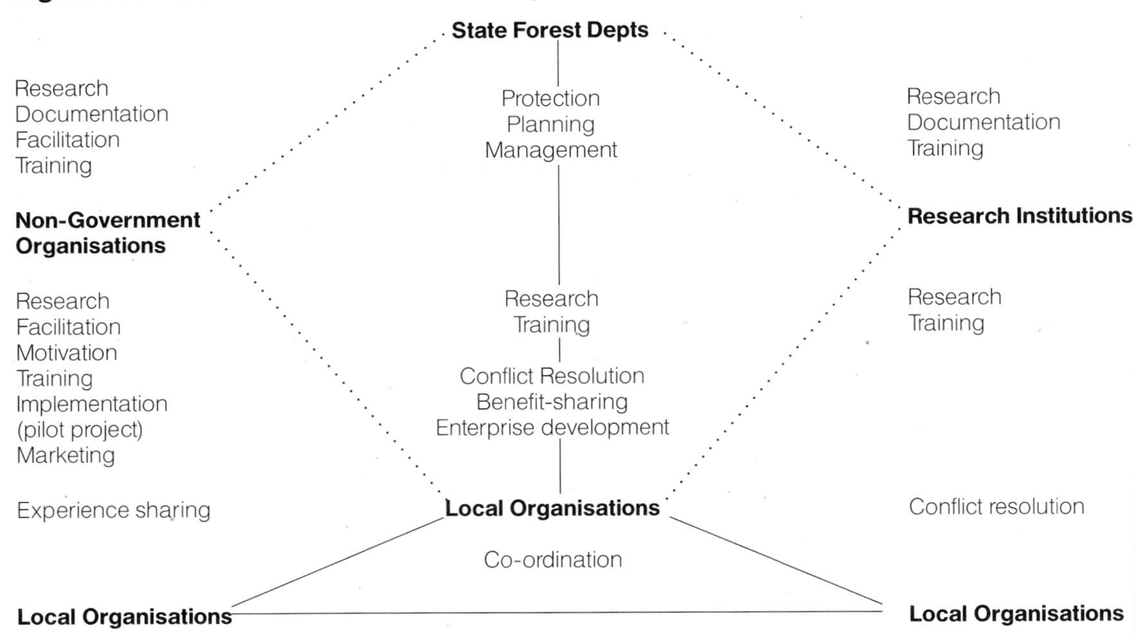

Source: Campbell *et al.*, 1994

more significant developmental impact, mainly at the local level; these are the implementational NGOs. In the main they have had a highly sectoral and location-specific focus, with a relatively restricted spread effect. Some have replaced the activities of the state, others have worked in association with state agencies, others have taken up an independent and critical function, working in both advocacy and implementational modes where they have attempted to influence policy based on experience.

Just as there can be no universal model for a successful local organisation, similarly it is hard to reproduce the success of one NGO in another environment with different actors. The role of key individuals, in positions of influence spanning the government and non-governmental sectors, cannot be underestimated; a strategy based on such people has been used successfully by several aid agencies. However, dependence on a particular 'stellar configuration' may also cause programmes to collapse when the individuals involved leave, lose favour or clash with incomers.

As notions of decentralisation gain impetus and the role of the state comes under closer scrutiny so do the role of NGOs and their relationship to the state. In forestry because of the vesting of land ownership in the state, this relationship has particular significance, and adds a further dimension and tension to NGO-GO relationships. Unlike in agriculture, where in general NGOs can play a direct role with independent producers who own their land and therefore have some degree of management control, in forestry the

Table 7.3 A typology of some of the main NGO and GO roles

Role	Features	Mode and level
1. NGO-GO partnerships within the policy – practice continuum	In joint NGO-GO projects, NGOs provide social organisational and delivery components; GOs provide technical inputs	*Collaborative*: each side relies on the other to contribute to agreed activities in accordance with perceived comparative advantages. In the absence of agreed inputs from one side or the other, the activity cannot fully succeed. Formal agreements are usually reached. e.g. VIKSAT and IBRAD running in-service training courses for the Forest Department
2.1 NGOs innovate; GO response varies 2.2 GOs innovate; NGOs respond	NGOs innovate – whether in technical, procedural, institutional or methodological ways – in the expectation that GOs will 'scale up' GOs innovate in a similar way to NGOs in the expectation that NGOs will spread the methods to new areas	*Incorporative*: GOs may disregard NGO innovations; if they do wish to adopt them, this may be through working together in the initial stages (i.e. as NGO teaches GO) before GO incorporates lessons from the NGO into its own actions. This also happens in reverse with NGOs in the position outlined for GOs
3. NGOs as 'networkers' among themselves and with GOs	NGOs establish fora in which ideas are exchanged among themselves and/or between NGOs and GOs	*Informative*: NGOs provide information on activities or on technologies to each other or to GOs, sometimes leading to coordination among projects or activities e.g. The JFM network
4. NGOs advocate; GO response varies	NGOs seek pro-poor administrative or legislative reform, or the full implementation of existing laws and procedures	*Conflictive*: NGOs seek to change GO practice through confrontation, lengthy negotiation or 'working from within' e.g. opposition to policy and legislation – the Draft Forest Act 1994

Source: adapted from Farrington *et al.* 1993

users of forests have at best only usufructuary rights and no management control. This control is held by Forest Departments. Therefore, if NGOs wish to work with forest users on state forest land, they have to assume an intermediary role between the user and the state. Intermediary organisations have an early and difficult catalytic role in facilitating linkages between the local people and the state. Several organisations in India, including VIKSAT and AKRSP (in Gujarat), MYRADA (in Karnataka) and TERI (in Haryana), to name but a few, have been relatively successful in this role. An example of the work of AKRSP is provided in Box 7.8.

Box 7.8 Development of Local Forest Management in the AKRSP Programmes in Gujarat

The experience of the Aga Khan Rural Support Programme (AKRSP), a non-governmental organisation working to facilitate participatory forestry in parts of Gujarat State of India, provides some insights into the essential role of a catalyst and facilitating organisation.

The primacy of a village-level organisation that takes on the responsibility for overall development and management of natural resources at the village level has been the cornerstone of AKRSP's approach to development. The 'Gram Vikas Mandal' (GVM) or a Village Development Association, comprising one male and one female representative from every household, was considered to be the ideal village organisation. The GVM is the forum for all village development-related issues. AKRSP's role has been to

- develop multifunctional local organisations
- facilitate a process of policy change at the government level based on field experience
- and working with government Forest Departments to help them to adapt their operations to respond to the requirements of participatory forestry

Source: Anil Shah, pers. comm.

Networks

As a precursor to more formal institutional arrangements, networks can be used as a means to facilitate exchange of experience and to begin to develop informal relationships between individuals and institutions working in a common arena. This is particularly the case where NGOs are either weak or have limited credibility with the target institution. Prior to the revolution in Nepal, there were few NGOs and therefore limited alternative avenues for local people to have a voice outside the formal government and political structures. Networks of user groups were encouraged by project and Forest Department staff. Networks also provide a forum for communication between clients who may be physically isolated or, because of the cross-disciplinary nature of the work, may not previously have been aware of the work of other institutions operating in disciplines ostensibly outside their frame of reference.

An important recent development in Nepal indicates the power of user-based networks. Here, user groups have come together as the Federation of Community Forestry Users of Nepal (FECOFUN) and have already had input into debate at the national level about the development of community forestry. The Federation is registered, with an *ad hoc* committee which publishes material for distribution to its members. It is now preparing to hold regional network meetings at which there will be male and female representatives from each of at least 25 district networks, selected from networks at the range-post level. Thus it is apparent that the fundamental building blocks of a nested organisation are already in place. Similar experiences are being developed in India under the auspices of the VIKSAT-supported programme in Gujarat, where local-level networks are being linked with State-level groups. Such examples of vertical and horizontal linking are encouraging pointers to the continued growth of a people-based movement for forest management.

There are no universal solutions to the complex environment encountered at the local level, and as such there is no one ideal institutional form. The principle behind the working group (discussed below), of drawing together different organisations with varying perspectives and skills, is a useful one and forms the basis for the development of different types of collaborative relationships. Whether the institutions are drawn together through a working group structure, or less formally and directly through networks, should depend on the particular context and analysis of the types of linkages necessary to help facilitate the development of local to national linkages.

As experience has shown in India (see Box 7.9), there are a variety of relationships and roles which can be played by institutions and individuals in the development of these new approaches to forestry. It is the careful meshing together of appropriate actors and information which provides an environment that allows change. Above all, it is essential that an enabling policy environment is created that allows the full development of incentives to encourage institutional change and collaboration.

Box 7.9 Partnerships in India

Experience gained in Haryana, West Bengal, and Gujarat indicates that outside resource teams can offer valuable assistance to the Forest Department (FD) and communities as facilitators in training, process documentation, ecological research, and communication flows – all of which are designed to capture field learning and rapidly inform programme management. Assisting senior FD officers in conducting periodic working group meetings to review programme progress and make strategic policy decisions based on diagnostic studies and field feedback has been a major contribution of these resource teams. In the final analysis, it is the commitment of dynamic and dedicated individuals inside FD institutions which make the valuable contribution to programme innovation. After identifying a small core group of leaders, an advisory committee could be established to plan strategies, secure time commitments and define responsibilities for programme development.

Another important issue involves the facilitative role of support institutions, such as NGOs, university researchers, outside specialists and other government agencies, in strengthening both FD and community capacity to implement JFM. Through diagnostic research, experienced NGOs and other researchers can assist the FD and communities in generating the knowledge to understand the ecological, institutional and economic parameters that need consideration in the design of sustainable forest management systems.

Source: SPWD, 1992

Institutional support groups for change

Since the late 1980s, the Ford Foundation has provided funds to form and maintain institutional support groups (usually called working groups) for participatory forestry, an institutional mechanism that brings to life the principles of the 'three-legged stool'. The nature of these groups has varied, depending on whether they were formed to assist government agencies, NGOs, or research organisations. They may perform a number of functions, including acting as a co-ordinating body, providing training or workshop opportunities, drafting policy, helping with problem definition and leveraging donors.

Because of the focus on forest agencies, working groups that assist the government have been the most common and prominent institutional support groups funded. Working groups appear to be effective in addressing problems that require:

- policy reform where action is needed on a large scale
- a learning process approach
- contributions from different disciplines or organisational perspectives
- momentum to overcome institutional resistance

The groups function to guide government policy by providing access to a variety of perspectives and serving as a coalition for change.

Working groups formalise the partnerships that underlie participatory forestry, and by regularly convening them at an operational level, Forest Departments can institutionalise the documentation, monitoring and internal learning from JFM. Working groups provide a forum for Forest Departments to involve NGOs and academics at the decision-making level, and to make policy and course corrections in a flexible manner (Campbell *et al.*, 1994). An example of the Gujarat State working group is provided to indicate representation and roles and responsibilities (Box 7.10).

Haryana has developed the working group idea to include working groups at division and range level in order to try to ensure that local groups are represented. However, it is still the case that village forest management groups are generally represented by members of intermediate organisations, and do not have direct representation. However, in some States, namely Orissa forest management groups are beginning to coalesce. In Nepal, the evolution of nested organisations is developing rapidly and provides valuable lessons for the institutionalisation of 'bottom-up' decision-making.

Nested organisations

If we return to the criteria for a robust organisation (discussed in Chapter 4), and consider more carefully what is meant by 'nested

Box 7.10 Gujarat State Working Group for JFM

Gujarat State Forest Department
Principal Chief Conservator of Forests
Chief Conservator of Forests, Development and Management
Chief Conservator of Forests, Social Forestry
Conservator of Forests
Deputy Conservator of Forests
Non-Governmental Organisations
Aga Khan Rural Support Programme
(pilot projects, facilitation)
VIKSAT
(research, training, pilot project, facilitation)
Sadguru Water and Development Foundation
(pilot project)
National Tree Growers Co-operative Society
(implementation, facilitation)
Academic Institution
Institute of Rural Management (Anand)
(research and process documentation)
Donor Institution (invitee)
Ford Foundation

enterprises', some interesting experiences from Nepal help to give operational guidance to a good idea.

With the growth of user groups across districts in Nepal, it became apparent to several projects that a more effective way of providing these groups with a voice was to encourage exchange and federation between the groups. These relationships were developed informally through a number of mechanisms including exchange visits and study tours to other community forestry areas (Box 7.11).

The potential for developing more formal linkages between local groups and forest planning operations was realised in the Koshi Hills area of Eastern Nepal, where workshops have been held to allow the full involvement of user groups in planning (Box 7.12).

Experience in Madhya Pradesh underlines the importance of linkage between decision-making fora. Interestingly, in Madhya Pradesh the vertical integration of JFM groups into Forest Department structures is broadened through linkage with other departments, but again there is no direct representation of JFM members on higher-level decision-making bodies. This is a significant difference between the Nepal and India examples; in Nepal representation of users is encouraged.

These informal structures have recently been formalised in Nepal with the adoption of range- post planning as a mechanism through which to link bottom-up planning with top-down strategic planning and budget flows (NACFP and CPFD, 1994). It provides for a standardised monitoring system that is not just responsive to donor requirements but which responds to the planning and implementational requirements of the Ministry and Department and is based

Box 7.11 Study Tours

Study tours have been used extensively by projects to expose user groups and Forest Department staff to new ideas. The major aim is to allow peer-group interaction and to foster innovation.

Where they are attached to projects, VSO volunteers have taken the initiative in facilitating these tours and have found that, in most cases, the most effective tours are those that are restricted to the district from which the participants come. The advantages are several:

- language is common
- ecology is similar
- social, cultural and political conditions are likely to be similar
- allows for follow-up visits by participants
- enhances networking possibilities between user groups
- allows for more in-depth focus on rural development issues and not just on community forestry
- easier to organise and less expensive
- reduces travel time and increases time with user groups
- shorter study tours increase the chances of women and other marginalised groups participating

There are some disadvantages:

- the tourism incentive disappears
- the sense of being part of a larger-scale initiative is lost
- the ability to learn from different project approaches is reduced

Overall the format of the study tour should be dependent on the objectives and the target group. In some cases, it may be appropriate to visit a number of areas outside the project to look at different approaches to similar problems.

Source: material draws on VSO volunteer reports

Box 7.12 Networking User Groups

In July 1991 the first user group networking workshop was held in Dhankuta district, the success of which led to several other such workshops in the project area. The workshops provided a forum for forest user group representatives and DFOs to engage in open discussions reaching agreement on issues related to planning and the implementation of programmes. The workshops increased the participants' conceptual understanding of the process of developing support systems and working strategies. Proceedings of the workshops were used as tools for planning DFO annual budgets and activities.

Although initially conducted by project staff with support from the District Forest Officer, the DFO increasingly took the leading role and found the workshops both challenging and rewarding. As the number of groups increased so did the number of workshops to be held.

Source: J.R. Gayfer, pers.comm.

on a wide range of experience from across the country (Wee and Jackson, 1993). It will also provide a much clearer negotiating base for planning resource allocation between the district development authorities and the sectoral agencies (Poudyal and Edwards, 1994). All new systems require refinement in the light of experience. However, there is one major problem to be resolved, namely the lack of institutionalised mechanisms to involve users in the planning process, although in the NUKCFP area, forest user group networking has begun to address this issue (see Box 7.13).

The power of user group networks in Nepal was demonstrated in 1993 when they held a national workshop at which representatives from user groups across the country exchanged experiences and provided recommendations for change in operations which were then presented at the 2nd National Community Forestry Workshop. This was the first time that the users had been able to express their views in a national forum where policy-makers were present.

Box 7.13 User Groups and Bottom-up Planning

Representatives from user group committees come together at a district or range user group networking workshop. The workshop serves a number of purposes: it provides an opportunity for district forest office staff to interact with groups, to explain the support available, to listen to problems; and for the groups to exchange experience and ideas. At these workshops the agreed annual plans for each user group will be presented and amalgamated to form the basis for the district or range plan, around which budgeting can be developed.

Source: Gayfer and Pokharel, 1993

7.5 The Impact on Project Process: From Blueprint to Process

Much has been written in the literature regarding project approaches (Korten, 1980; Rondinelli, 1983). All writers contest the use of so-called 'blueprint'-style approaches and advocate the need for a learning process where projects and institutions evolve; based on analysis of experience and the adaptation of practice accordingly (see Boxes 7.14 and 7.15 and ODA, 1995 for detailed discussion). In this way the style of the project is in accordance with the form and structure of the new institution.

In complex systems with inter-related problems there are no defined solutions, and therefore it is necessary to devise strategies and a sequence to address problems over time. Process projects establish mechanisms that permit revision and development of project outputs as experience is gained through implementation. In concrete terms this means that projects are designed with indicative frameworks for budgeting and planning and the finalisation of both inputs and outputs is determined by the project partners. The

Box 7.14 The Blueprint Approach

Features of a blueprint project:

- excessive size and complexity, based on the unproven assumption that there are always economies of scale
- cost-based monitoring, excessive preoccupation with spending agreed sums of finance, partially because staff are assessed in large measure by their ability to spend up to budget, but more fundamentally because high levels of expenditure on administration and project supervision can only be justified by high levels of project spending
- limited time-horizons, imposed on projects by the often arbitrary project-life
- an inflexible and prescriptive approach, leaving little scope for redesigning the project and 'learning-by-doing'
- unilateralism – so-called project beneficiaries are given little scope for influencing project design and implementationa bias towards measurable goals on the grounds of bureaucratic convenience
- little or no emphasis given to the development of institutional capacity to support project-induced initiatives after the project is withdrawn

Source: adapted from Moris and Copestake, 1993

Box 7.15 Process Projects

Features of a process project:

- smaller, longer, more flexible projects with the emphasis on a pilot-scale learning stage
- emphasis on development of a conducive policy and institutional framework for the implementation of project process beyond the pilot stage
- qualitative milestones (markers of change and progress along an agreed pathway) open to review and to change by project partners
- project partners are involved at all stages in the project cycle, including project design and evaluation
- projects are designed with 'in-built flexibility so that local institutions and groups learn to develop their capacity to identify, plan, manage and replicate projects'. (Eyben, 1991)

emphasis, of necessity, is placed on institutional development, since these types of projects require highly decentralised and flexible structures. Both aspects, as has been discussed above, are rarely present in South Asian Forest Departments.

One of the most difficult challenges facing any donor organisation is how to address and change the culture of a bureaucracy that currently acts in a way that is inimical to the achievement of the objectives of the development project. Experience to date in the forestry sector provides some guidance:

- Ownership of the project process (ideas, implementation patterns, etc.) must be developed by the implementing organisation.

- There must be development of pilot-level approaches with in-built linkages to allow development beyond the pilot-level.

- It is insufficient just to develop consensus about the purpose of the project, since in most cases understanding is only really demonstrated when the project partners are able to plan and implement activities that will contribute to achievement of the purpose. (Experience has shown that what was considered to be mutual understanding rapidly deteriorates to mutual incomprehension when it becomes clear to the implementing agency what types of activities the donor considers to be essential to implement the project successfully).

- There cannot be too much joint planning, i.e. the implications of the 'vision' for institutional change need to be jointly articulated.

- The project 'vision' must be owned by all levels of the implementing agency (i.e. the situation must be avoided where only senior management know where the project may be going to, and where junior staff are merely responding to orders).

- A clear understanding of the hidden institution must be developed by the donor organisation, since denial of its existence has had serious negative consequences for project success.

- Project managers must act as catalysts and facilitators and not as prescribers of action.

- Clear identification of client groups, and methods for their inclusion in the project process, must be developed. Just as project 'vision' should be owned by the implementing agency, it should also be owned by local people.

- Decentralisation without devolution of control will not work.

- Transparency and accountability at all levels (within the village and within the bureaucracy) are essential.

- Institutional arrangements should respond to the local conditions.

- Local-level planning fora should be linked into existing or new planning structures to provide local people with a real voice.

- Strong local-level organisations need to be developed that can network and federate amongst themselves, and are able to link with other organisations as is appropriate.

- Adequate incentives for involvement in the process should be ensured, incentives which will be sustained beyond the life of the project.

- Policy frameworks must be enabling and not prescriptive, i.e. permissive of idiosyncrasies.

- Property rights, in many situations, need to be clearly articulated and strengthened at the local level.

- There is a need for donor reorientation to ensure that flexibility and responsiveness are built into their bureaucratic structures.

- A strong constituency for the project should be developed. If there is no demand from local people for participatory forestry projects, there can be little point in continuing. The lack of demand probably indicates that the project has been poorly developed without ensuring that the above steps have been followed.

In this search for mechanisms to effect institutional change, projects and donors are being drawn into debates that require generic lessons to be drawn from the diversity of experience: the need to generalise from idiosyncrasies, and to provide principles upon which future action can be predicated. As Chapter 8 highlights, much of the evidence needed to develop these principles and understanding is still lacking, and should be collected as part of the participatory forestry process to ensure that change is sustained beyond the life of these programmes.

Tools are emerging that help in the monitoring and assessment of process change, and they are being refined as experience is gained. In India, based on earlier work in the Philippines, process documentation has been used to some effect in Haryana to help inform project implementers of the impact of participatory forestry (Box 7.16).

7.6 The Implications for Policy and Practice

The future being contemplated by the public sector institutions is dominated by questions of accountability and responsibility. The process of change and its facilitation are the key questions facing donors wishing to support public sector reform; participatory forestry has provided one of the entry points for the reform process to start.

Since the underlying premise of all these changes is that solutions must be derived from an understanding of the problem context, the starting point for the forest sector is the forest and an assessment of the stakeholders, and their interests, and methods of reconciling competing interests and defining acceptable management objectives. As was discussed in Chapter 4, this is the case whether the forest is to be managed for conservation, economic or local livelihood objectives (or perhaps some combination of all three).

Such an approach has implications for each level of the Forest Department. For planners their role is to provide strategic planning frameworks (i.e. simple guidelines for assigning priorities to particular forest areas – or groupings of areas in the case of Protected Area Networks). However, in order for planners to be able to

Box 7.16 What is Process Documentation?

Process documentation (PD) records the process of interaction with and between all parties involved in a programme of activities. In any new and innovative programme for which no proven blueprint for implementation exists, PD can be an invaluable tool for collective learning and ensuring practices remain relevant. The following example draws on Madhu Sarin's work in the Shiwalik Hills of Haryana State.

Essential features of PD

1. Any interactions (discussions, observations) made by a project team member when visiting a village are recorded.
2. This ensures that continuity of learning is maintained even if team members change.
3. Allows a rapid sharing of information.
4. Ensures that key issues are highlighted.
5. Relevant action is taken at the appropriate level to resolve problems.
6. Information is nested – i.e. detailed field notes are kept and referred to by field-level teams, and a summary of action points not resolvable at that level is made by these teams and passed up the system for action. In this way sometimes unnecessary detail is avoided, and clear responsibility assigned for particular activities.

In order to ensure that PD is an effective representation of what is happening the following activities need to occur:

1. Prior to going for a village or working group meeting, the main issues to be discussed should be identified by the project team.
2. It is essential that the facilitating team develop the ability to listen and learn, and to throw away an agenda if it appears to be inappropriate.
3. Documentation of change is not sufficient, it should lead to a change in activities if necessary. For example, an anomalous pricing structure may lead to a disincentive for local involvement in JFM, mechanisms should be in place to allow this information to be fed to the level at which this anomaly can be addressed.
4. Above all, the effective use of PD requires a bureaucratic structure that is willing and able to respond flexibly.

Source: M. Sarin pers. comm.

develop such guidelines, detailed contextual understanding is a pre-requisite, as is an enabling organisational structure that allows lower-level staff authority and freedom to make decisions on the basis of local knowledge. Similarly, for policy-makers their role is to respond to local complexity and diversity with the development of policy that provides an 'enabling' framework to allow local decision-making, rather than a prescriptive framework that reduces complexity to a few simple and inappropriate rules (Table 7.4).

All activities flow from this understanding and the ability to plan according to what is required at the local level, together with an accommodation of more strategic needs. At each level of planning problems should be resolved; where there are clear policy implications or blockages to implementation due to organisational constraints higher up the system, these should be referred upwards.

Table 7.4 The planning framework

Level	Outcome
State Working Groups [Secretary of Ministry, other relevant secretaries or heads of depts, NGOs, CCFs, (ideally local reps), CFs, research institutions/ universities, donors	Policy change Strategic research needs Strategic planning framework Bureaucratic reform Inter-organisational co-ordination Identification of external resource people to support programme development
Circle Working Groups [CF, DFOs, NGOs, reps of local forest management organisations, universities/ research institutions in circle or nearby, donor rep (invited)]	Issues for policy reform Inter-organisational co-ordination Inter-circle co-ordination and exchange of experience Circle planning Management objective prioritisation Guidelines for 'best practice' Adaptive research questions Identification of resources to carry out research Joint implementation of research Identification of training requirements and local agencies to provide skills, emerging from implementational experience
Division Working Groups [DFO, ACFs, RFOs, other line agencies, NGOs, reps of local forest management organisations, universities/ research institutions in division]	Co-ordination of experience exchange between ranges Objective prioritisation on basis of local-level planning Co-ordination of action research across a series of ranges Co-ordination with other line agencies and NGOs Monitoring and review of experience, redesign of programme
Range Working Groups [RFO, Foresters, FG, reps of local forest management organisations, NGOs]	Conflict resolution Experimentation with new silvicultural approaches Action research Operational change in the light of experience Identification of policy and legislative constraints Planning for range-level budgeting Networking between local forest management organisations Identification of resource needs and suppliers
Local Forest Management Organisation [All users – male and female]	Planning of objectives for management of forest Annual work plan for management of forest to form basis of range-level plan Identification of action research issues Management of forest resource Resolution of inter- and intra-group conflict

7.7 Discussion

> Any single, comprehensive set of formal laws intended to govern a large expanse of territory and diverse ecological niches is bound to fail in many of the habitats where it is supposed to be applied (Ostrom, 1994)

What, then, are the policy and implementational implications of this statement? It leads us to a fashioning of policies which provide a general framework to allow local flexibility to be accommodated. It requires responsive organisations in which individuals have the freedom to respond to need and to identify resources to support a particular approach. It also implies a highly risky environment where the safety of 'norms' is abandoned for the uncertainty of 'anything goes', but within tightly monitorable frameworks. It calls for a world bounded by strict rules of accountability and transparency at every level.

Recent experience has shown that the expansion of participatory forestry brings its own challenges. As Forest Departments become eager to expand programmes, they are learning that there is not always the time or the capability to train staff adequately. Also, where funds need to be rapidly disbursed to a large of number of sites, there is often a tendency for decision-making to become more centralised, which tends to lessen the agency's capacity to respond to heterogeneous conditions. In addition, experience has shown that the time and effort required to organise and sustain activities in a single village have often been extremely demanding on the resources of the implementing agency, although some of these costs can be considered one-time expenses associated with establishing the programme. How, then, can participatory forestry programmes expand without sacrificing their essential character, namely the capacity to respond to diverse social and ecological conditions and to give adequate attention to implementation?

Diversity is both a challenge and an opportunity. However, to bureaucracies used to the implementation of wide-scale models, diversity is considered to be a problem whose solutions often lie beyond the competence of highly structured organisations. The development of new institutional arrangements and partnerships has to a certain degree, begun to accommodate local diversity. Thus in India and to a more limited extent in Nepal, positive relationships have been constructed between locally based NGOs and government departments. The NGOs have provided the flexibility and responsiveness required to accommodate social diversity, and the government staff have been able to work alongside NGO partners to provide the technical response to ecological diversity. Although NGOs have provided additional extension outreach in many countries, it is also the case that it is not sufficient or indeed adequate to rely on NGOs alone to provide the interface between local people and government departments. In India, for example, the emphasis on the development of joint partnerships between local people and Forest Departments has been of central importance. It ensures that Forest Department staff also learn how to become more responsive to local needs. This is the case particularly in Nepal where the absence of intermediary NGOs has forced the development of an active interface between local people and government staff.

Above all, what has become apparent from this analysis of experience in India and Nepal is that there is no one solution, but rather a continuum of institutional arrangements and levels of abstraction at which they operate. There are as yet no answers to the many questions of when to encourage particular types of partnerships and when the state should retain control. But now that the questions are being more closely defined, it will be easier to begin to collect information to clarify empirical evidence and to begin to define generic principles for both policy and practice.

To end this discussion as we began with another quote from Ostrom (1994), the challenge for the next decade is:

> the *match* of institutions to the physical, biological, and cultural environments in which they are located that will enable institutions (and the resources to which they relate) to survive into the twenty-first century.

Notes

1. This section draws on Wollenberg and Hobley (1994).
2. This Guide does not attempt to review the major works analysing the role of NGOs. For further information the reader is referred to the important series of books by Farrington *et al.* 1993; Carroll, 1992.

8 A New Pragmatic Forestry or Another Development Bandwagon?

Mary Hobley and Eva Wollenberg

8.1 Introduction

In thinking about the future of participatory forestry programmes, several factors can be taken into account, including the stage of development of the individual programme, changing country contexts and new trends and levels of understanding in the area of people, rights and resources. This Study Guide has reviewed the historical development of the forest sectors in both India and Nepal and has focused on recent changes in management arrangements. The analysis has been placed in the context of global changes in forestry, as well as the individual country contexts of public sector reform and divestment. This final chapter looks at the evidence collected, and questions whether these shifts amount to a new paradigm or just a marginal addition to existing practice, an addition that will have limited impacts beyond the life of donor projects and government programmes.

8.2 Major Issues to be Addressed

While participatory forestry programmes have made important advances, they have also faced significant challenges. In discussions with forestry officials, NGO workers, and villagers, as well as with donors, several questions have been raised which the preceding chapters have considered but to which there are still many unanswered aspects. These include questions of equity, empowerment, income generation and the long-term role of Forest Departments as facilitators of social change. The following sections look at these major issues in greater detail, and suggest ways forward in the search for answers.

While accepting that little systematic information is available about impacts at the village level, we shall nevertheless turn our attention to examining how effective programmes have been in bringing about meaningful institutional change. Mechanisms for sustaining improved forest regeneration or sharing of rights and responsibilities of forest management are still not clearly articulated, but without such mechanisms it is questionable whether these systems can survive. These mechanisms include support to or development of local organisations, contractual agreements, multi-scale and inter-organisational co-ordinating bodies, new skills and technical knowledge. Changes in tenure and rights of access to forests and forest products are relatively fragile, recent introductions. It must be questioned whether approaches such as joint forest management will be sustained into the long term or are merely convenient tools to fit the current political and developmental context that will be extinguished as soon as there is a change in political will. As was discussed in Chapters 4 and 5, the patchy evidence to date does not provide clear support for the achievement of the original objective of participatory forestry – that of enhancing the forest-based livelihoods of local people.

Who benefits?

How effectively are the participatory forestry programmes reaching villagers?

Since many of the participatory forestry programmes emphasise the importance of facilitating institutional change and working through other organisations, there has been a natural tendency to assess the programmes in terms of the changes in institutions rather than the impacts on villagers' lives. Thus relatively little is known about how villagers' well-being improves with, for example, the handover of forests under community forestry or joint forest management. This information is lacking in part because of inadequate resources or systems for monitoring. In some sites it may also be too early to see the real impact on villagers' lives. Where monitoring does not occur, rapid evaluations of the programmes are sometimes conducted. These reviews typically do not permit rigorous assessment at the village level. Such experience suggests that there is an urgent need for more systematic monitoring to provide a basis for programme learning as well as to inform policy-makers about the impacts of different policy instruments. Attention needs to be given to the question of how to design simple, affordable monitoring methods that will channel information to key people in a stimulating way. In particular, there is an urgent need to develop local monitoring systems implemented by the users and managers of forest resources. As has been noted in recent meetings of the Joint Forest Management Network, part of the difficulty in tracking

Box 8.1 The Cynic's View of Participation

Participation is:

- No longer perceived as a threat. Governments have learned to control the risks of 'unruly' participation and are interested in the increased productivity at low cost. Development policies tend themselves to create induced and addictive needs in 'target populations'. Thus their 'participation' is used to secure general support for the same needs.
- Politically attractive as a slogan. Here governments have learned to control it, and they benefit through prominent displays of participatory intentions.
- Economically appealing. In conditions of economic trouble, passing on the costs to the poor can be achieved in the name of participation and its corollary – self-help. It is a means of mobilising new resources (mainly in the form of labour) for activities recently funded by government.
- An instrument of increased effectiveness and investment. Grassroots organisations become the infrastructure through which investment is made, or they help provide the 'human software' that makes other kinds of investment work.
- A fund-raising device. Governments at both the giving and the receiving ends via the use of the participatory discourse may mobilise other avenues of funds through NGO fund-raising programmes.

benefits to villagers is also that many of the most important changes are difficult to observe and measure, for example empowerment, security and voice. This implies the need for methods that are also creative and take these less tangible kinds of benefits into account. The urgency of providing such methods is immediate, if we are to be able to answer some of the questions raised in the preceding chapters.

Having better information about village impacts is ultimately the only definitive way to assess whether the programme is meeting its objectives and is worth continuing. Success can be counted as only partial if it occurs at the institutional level. As Rahnema (1992) points out, people's self-interest in these institutions is also at play in the adoption of participatory programmes (see Box 8.1).

Many aspects of the cynic's view may be relevant. However, until there is evidence to support or refute these suppositions, it is difficult to determine with any certainty the underlying motivations for governmental involvement in participatory approaches.

Analysis of the 'real' costs and benefits of participatory forestry

Chief amongst the questions still to be answered is how great are the real costs and benefits of participation and how are they distributed amongst the various actors. These issues need to be tested empirically across India and Nepal. In particular, the unquestioning assumption that participatory forestry is of benefit to local people must be rigorously tested. Some of the key areas to be addressed include assessments of:

- the labour costs incurred in: forest protection; collection of forest products from more distant unprotected forests; attendance at meetings; increased administration, etc.
- how access to decision-making and resources changes
- the effect of changes in village/local government relationships
- the effect on the flow of other resources to the village: employment, other development initiatives, etc.
- the control over marketing structures
- the amount and distribution of benefits from timber and non-wood forest products, especially in comparison with other income flows
- the increase in volume of productive biomass from protected forests
- the species mix of biomass produced and which interest groups are benefited and which disadvantaged

Such questioning also needs to be directed towards the implementing institution in order to assess the costs and benefits of participatory approaches to Forest Department staff. Incentive and budgetary structures, as discussed in Chapter 7, often run counter to the working practices necessary under a participatory approach, and staff may find that there are few tangible rewards in working to develop participatory forestry systems. If there are few incentives under current institutional structures for following a participatory approach, such programmes will not continue to function into the future once the catalyst is withdrawn.

In some cases, programmes may have costs not taken into consideration in the calculation of benefits. For example, most JFM programmes in India do not take into account the costs of timber harvesting and other operations, yet if these were deducted from the benefits available to the community, the percentage of benefits would be significantly reduced. While activities such as felling timber and guarding plantations are fairly straightforward costs to itemise, should the cost to the community of their organisation also be taken into account? Anyone who has worked in a participatory project knows the extent to which farmers bear significant time costs to attend meetings and take decisions. Just as there is a need to be able to measure some of the less tangible benefits, the invisible costs need to be identified and calculated.

There is as yet no one satisfactory means, whether by contingency valuation, shadow pricing or time allocation, of assigning a single common unit to all forest benefits and costs. Although marketed products can be assigned a monetary value, other products do not necessarily provide the same level of perceived benefit to all individuals. Some households will barter the good, others will consume it. Similarly, labour will be valued differently by households depending on their opportunity costs. This suggests that it may be unrealistic to come up with a single measure of costs and benefits, and systems should develop multi-dimensional scales for these assessments, rather than conventional input-output monitoring systems.

In summary, if one of the key aims of participatory forestry is to improve the benefits to local people, it is important to be able to observe these benefits and to have in place means for ensuring the allocation of project benefits over the long term. There is a need to:

- measure programme impacts on people, not just institutions

- develop institutional tools that better articulate how benefits will be generated and distributed

- account for the less visible, in-kind or intangible costs and benefits of projects and

- disaggregate social groups in order to have a more sophisticated understanding of who is bearing the benefits and who the costs, and how project goods should be distributed.

If participatory forestry is to move forward, programmes will have to devote more attention to not only generating benefits but also more closely observing what those benefits are and who is receiving them. In the next three sections we discuss empowerment, competition and income generation as other emerging areas requiring more attention in the 'next generation' of participatory forestry programmes.

How can participatory forestry programmes ensure an equitable or targeted distribution of benefits within communities?

The experience reviewed in this Guide has also indicated that little is known about *which* villagers within a community benefit and what the social dynamics underlying this distribution are. Because the more powerful members of a community may control the benefits of a project, it is important to know how groups specifically disadvantaged by class, caste, gender, or ethnic group can be assured of their share. Even where projects have been judged 'successful', there may be instances where participatory forestry programmes have unwittingly exacerbated difficult conditions for some groups. In the same way that Forest Departments and communities share control over resources, participatory forestry approaches could explore mechanisms by which advantaged and disadvantaged rural households could co-manage resources and share the resulting benefits.

The need to disaggregate social impacts calls for careful use of the term 'community' and more attention to the intra-group relations and constituencies, including the variety of locally relevant social groups such as user groups, women's groups, youth groups, village forest committees, indigenous forest management organisations, schools and user group federations. As was demonstrated in Chapter 4, there is a large array of both formal and non-formal mechanisms through which to manage resources. Again, the challenge lies in the identification of appropriate structures for each situation.

Empowerment

Empowerment is defined here as a process of increasing control and influence over decisions and is achieved by a number of means. Efforts to increase villagers' security of land tenure, enhance their incomes and improve their participation in land-use decisions are part of an explicit strategy of participatory forestry programmes to transform the economic or political power relations between communities and forest bureaucracies. Similarly, encouraging forest bureaucracies to be more responsive to community concerns is an explicit means of increasing villagers' control vis-à-vis government. The two strategies are complementary: local groups' control in forest management can be strengthened or the control of the bureaucracy can be lessened. The unequal relationship can also be modified by creating mechanisms for the mediation of competing interests between government and local communities or creating incentives for co-operation based on common interests. Other means include building capacity among local groups, strengthening NGOs as intermediaries representing villagers' collective voice, and providing legal education or assistance.

This raises a further issue about the extent to which local organisational development will continue without the presence of an external facilitator. In Nepal, for example, the emergence of user group federations and exchange visits among user groups is beginning to address some of these questions. Although external facilitators are still necessary, in particular to help resolve conflicts, increasingly these user groups are gaining power and internal strength through their federations. Some new horizontal relationships have been formed, and groups of people (in particular, women) previously excluded from the decision-making process are gaining greater access to it (as discussed in Chapters 4 and 7). These alliances are a form of empowerment and allow local people to exercise their rights to self-determination. Environmental interest groups play a key role in many countries in supporting participatory forestry programmes and in some cases exerting direct political influence.

Accompanying the development of federations and the formalisation of group activities, comes a second generation of problems. This was highlighted by experience from both India and Nepal, where increasingly demands are being made for payment (in cash or kind) for participation in these groups. The increasing formalisation of the groups and meetings and the regularity of their occurrence may make them less accessible to women, who cannot commit the necessary amounts of time; and as the groups form into federations there are increasing signs of their becoming politicised, with the accompanying risk of their being used by political parties. Thus these new organisations may also create their own internal power dynamics with competing self-interests that may incur unexpected costs. These questions require further analysis.

In order to increase the understanding of the different approaches used for empowerment in participatory forestry, it would be helpful to know more about the impacts of these strategies at the village level. Questions to assess empowerment include:

- How do these different tools of empowerment (e.g. land tenure, income generation, strengthening cultural identities) enhance villagers' control and influence over their livelihoods?

- Which members of a community tend to be empowered through participatory forestry programmes?

- Does empowerment of some occur at the expense of others? As some members of a community gain access and control over resources, does this displace others from access to these or other resources?

- What are the opportunities for a more demand-driven approach to participatory forestry?

As experience has already shown, there may be limits to the degree of empowerment achievable at the local level whilst working entirely through Forest Departments. It may be the case that it is not possible entirely to transform the working relationships between bureaucracies and the citizenry. A critical question, therefore, still to be explored through the programmes is:

- The degree to which the state is prepared to delegate power to its citizens for the control of natural resources? and the degree to which this should be promoted through the forest sector separately from other natural resource sector initiatives?

Resolving competition

Participatory forestry programmes are in part a response to competition among diverse interests for increasingly scarce forest resources. While competition is not inherently undesirable, it can lead to delays, tensions or even violence and can block access for some people to the resources on which their livelihoods depend. The major interests with a stake in forest management on public lands include local communities, the private sector (including forest industry) and the state. Competition may occur within as well as among these constituencies.

One example of lack of clarity about the underlying interests is the notion of 'public interest'. Much of the conservation work in participatory forestry is promoted by governments (or the international community) on the basis that it serves the public interest. Yet these groups may be using the term to justify meeting their own ends. If different government bodies have political biases towards meeting the needs of one constituency over another, should government alone be left to define the public interest (although it is often assumed that this is its role)? Similarly, should international environmental groups with a specific agenda define the global public interest? The concept of public interest may provide a rallying point around which diverse governmental and non-governmental parties can be organised to express their views and jointly determine public interest.

At least four strategies were identified by the Ford Foundation as potential mechanisms to resolve constructively the conflict associated with competition (Wollenberg and Hobley, 1994). Support may be given to:

- Widen, restructure, inform or clarify 'the debate' surrounding the conflict. This may include identifying areas of common ground. The media and universities can play an important role here in providing new information.

- Facilitate contact among the parties to develop interpersonal ties. Such ties could help ease communication and encourage understanding and compromise.

- Enable a decision-making process to which all parties can agree. People are generally more willing to make compromises if they feel satisfied with the process. Cultural sensitivities need to be employed in determining whether 'democratic', consensual or other forms of decision-making are appropriate. It may also be helpful to encourage the representation of a wide range of perspectives to diversify the stands taken on issues and discourage polarisation.

- Enable the disadvantaged parties to organise, to mobilise resources and to gain the information necessary to negotiate their position. If support is given to a specific group, the basis for doing so needs to be very clear, i.e., in what ways is this group disadvantaged?

In general, competition can be better managed if these strategies are implemented early in the process, before conflict escalates and antagonistic relationships become entrenched.

Incomes

The rapidly changing context of poverty, in India in particular, may demand new approaches to resource management and economic development. As people living in remote areas have gained increasing access to markets and participate more in the commercial economy, at least three issues have become apparent:

- increased reliance on purchased goods and services has generated a need for cash incomes;

- the attractiveness of applying labour and capital to enterprises with higher returns than those traditionally achieved from forest use has lessened villagers' (and local governments') economic incentives to engage in participatory forestry activities and conserve forests; and

- better access to markets and an expanding range of products valued in the marketplace encourage rapid depletion of forest and land resources, unless appropriate sanctions are in place.

Participatory forestry can help to stabilise incomes, but there is little documentation to show that it has even provided significant additional income to village households. How, then can programmes help generate cash incomes and still meet the conservation and sustainability aims of participatory forestry? One key is to increase the range of villagers' income-generation options. Another consideration is to recognise that households use incomes to meet multiple objectives, including economic security, status, satisfying consumption needs and future investment. The amount of income generated might not accurately reflect its value in the household economy, although it is usually a good indicator.

People in forest areas face special economic conditions that affect their income choices. While they have access to forest products, they are often far from the infrastructure that facilitates adding value to such products and the competitive markets that enable them to capture the product benefits. Options for increasing forest dwellers' income in ways that are compatible with conservation include forest-based and non-forest-based choices.

One option that has developed with the increasing interest in

NTFPs is to increase the value of forest resources. This option includes identifying and creating markets for non-timber forest products; capturing added value through local processing; decreasing transportation and middleman costs; improving access to markets; creating values consonant with environmental amenities, such as ecotourism and parks from which user fees or wages from park-related employment could be acquired; and institutionalising royalties for genetic prospecting. Among conservationists, enhancing the value of the forest is seen as a means of making it more attractive to maintain than to destroy. This objective is not easily compatible with the aim of maximising the short-term economic returns from the forest. Thus far, it has been difficult to demonstrate that income-generating uses of the forest do not ultimately increase the incentives for further extraction.

Other options for increasing 'environmentally compatible' incomes encourage villagers to increase the productivity of their *non-forest-based* activities. These activities do not necessarily depend on forest land or resources. Depending on the scale of activities, these options might go beyond the scope of participatory forestry programmes and demand their own set of focused and coherent programme interventions. These additional options are:

- To increase the intensity or economic value of agricultural production on villagers' farm lands.

- To create capacity among villagers to undertake non-land-based enterprises. Such capacity would include identifying their needs for credit, skills and organisational development, technological inputs, raw materials and marketing, and might involve developing closer links between villagers and the private commercial sector.

- To develop villagers' skills in order to enable them to participate with higher levels of compensation in the wage economy.

- To enhance villagers' assets in order to give them the financial base to pursue independently different livelihood choices. For example, to determine the effects of different usufructuary land tenure arrangements on villagers' access to credit.

- To protect the security of existing or promised incomes. To create policy measures that protect villagers' income security, for example their access to existing forest-based incomes such as the share of income from harvesting timber.

Using any of these options, concerns about the amount of income (is it sufficient to lift villagers out of poverty?) and the equity of income distribution (who is earning and who is possibly losing out?) need to be given close attention.

Other concerns have been highlighted in this Guide, particularly in Chapter 4, where consideration was given to the impact on women of increasing the values and marketing of non-timber forest products, and whether it would lead to women losing control to men over the use and sale of non-timber forest products with their increased commercialisation.

Other questions that need to be addressed include:

- How are the new incomes being spent? Are they being reinvested in the community and resource base?

- Are the people who earn the incomes the same group that is making the management decisions? If not, will the investments required to manage natural resources sustainably be made?

- Are villagers' ranges of livelihood options increasing or decreasing?

Poor villagers' survival strategies are often based on maintaining several income sources at once; any reduction in the range of choices and the capacity to pursue multiple income streams might therefore cause more problems than it solves. To understand these strategies better, more needs to be ascertained about existing sources of income from, for example, urban migration, sales of surplus produce, occasional local wage labour and remittances from relatives.

As this and other directions such as villager empowerment and resolving competition are explored, one issue to revisit periodically is the extent to which the initiative should be incorporated into existing participatory forestry programmes or be taken up by other agencies and organisations that explicitly address income generation, community organisation or conflict resolution, thus returning to the question of whether forestry is the only (or necessarily the best) vehicle for generating these changes.

8.3 The Ideas Cycle: Phases in the Development of JFM

As a result of the analysis of the evolution of the participatory forestry programmes in Nepal and India, a distinctive cycle appears to be emerging. In this example, drawn from an analysis of the JFM programme in India carried out during a review of Ford Foundation's support to forestry in 1992, some interesting patterns have become apparent which help to guide analysis of the impact of the programme (see also Jackson, 1993 for a discussion of similar patterns in Nepal). They also help to provide a framework through which to consider future strategies for activities in the natural resource sector.

The informal innovation phase

This phase could be said to have begun in West Bengal with the pioneering work of Ajit Bannerjee in Arabari. It had its foundations, however, in the already long-established support to social forestry-type programmes in several States in India, and the indigenous forest management systems of Orissa (Kant *et al.*, 1991). Innovation in West Bengal, together with experiences from Haryana, Gujarat and Orissa, did lead to significant policy changes in the 1980s.

Promotional phase

This phase began in 1988 with the initiation of a new forest policy that highlighted the importance of people's participation, and provided the foundation for the passing of the JFM Resolution in 1990. This series of policy changes provided the appropriate enabling environment for practice to develop. New partnerships between key actors in the forest sector, particularly in West Bengal, were fostered. Some of these were made possible through a series of grants funded by the Ford Foundation.

Joint Forest Management has gained widespread acceptance and funding from the international donor community over the last few years. A clearly defined strategy succeeded in bringing the experiences gained in West Bengal and other States in India, to the attention of policy-makers, and resulted in the passing of the central government resolution in support of JFM.

Although this indicates a major change in government policy and commitment, it is only one small part of the process of institutionalisation of the underlying principles of JFM. This was recognised by the Ford Foundation and a series of support mechanisms were established, including the nationwide research networks (discussed in Chapter 7).

There are, however, inherent problems in sustaining the promotional phase for too long: the nature of promotion often leads to an overemphasis on the positive aspects of the programme with little critical analysis. Although at the outset it was important to be able to persuade key actors that JFM had many merits that outweighed those of the more traditional approaches, it was also important to temper this enthusiasm with long-term strategy and the building of capacity to implement such policies. During the promotional phase the proponent of a particular programme or idea may wish to promote the cause of disadvantaged people. In the case of joint forest management, the proponent feels that JFM may be one means to accomplish this end. In the effort to help local people, a new language and rhetoric take over, with the result that the distinction between 'the right cause' and the 'right means' often becomes blurred, with perhaps undesired and unintended consequences for local people, as was described in Chapter 5.

In this case the promotional phase has led to massive donor interest and funding for JFM, which has, in its turn, imposed pressures on the forestry bureaucracy for expansion of the programme which are difficult to sustain: in many ways the idea was promoted ahead of the capacity to implement it (as was discussed in Chapter 7; see also Kant *et al.*, 1991). Apart from the lack of institutional capacity, the technical skills to develop different silvicultural systems and to fulfil the varied objectives of management are also underdeveloped, although interesting new approaches are being tried (Chapter 6).

The outcome of this phase could be summarised as follows:

- change in policy at centre and State-level
- limited development of institutional capacity
- widespread international awareness
- massive increase in donor funding.

Consolidation and analytical phase

The promotional phase has now been replaced by a careful evaluation of the options for future action. There is also some emergent experience that can be assessed to provide indicators of problems and opportunities. Many have called for a period of consolidation before there is any further expansion of the programme beyond the capacity of institutions to support it. This implies a need for understanding what works, where and why. However, as demand for the programme has been fostered within local communities, NGOs and Forest Departments, it is now difficult to stop the expansion. Indeed, if the learning processes are in place, expansion can lead to increased understanding of the limits of the approaches under different conditions.

The need for consolidation and reflection is further underlined by the diversity of opinion about what the objectives should be for JFM. What is JFM? – a simple but revealing question which, in a meeting of Ford Foundation supported organisations in Delhi in July 1992, produced as many definitions as participants. Such diversity represents an opportunity to bring together new ideas and approaches; it also becomes a problem when there is no consensus on the expected outcomes. A number of outputs could be expected; for example JFM could

- give greater and regularised access for local people to forest resources and decision-making about forest management
- ensure that those who are most dependent on forest resources, i.e.

women and poorer people (both male and female), have increased access to resources and decision-making

- reassert Forest Department ownership over forest land and stop encroachment
- retain or increase biodiversity
- enable a restructuring of bureaucracy to make it more responsive to the needs of various actors in the use of resources.

The preceding chapters have provided partial answers to some of these issues but much more evidence needs to be collected from different environments to enable any real confidence to be attached to them. In the initial stages, the fact that JFM provides different benefits to a diverse group of people and institutions is one of its strengths. However, as JFM gains greater acceptance, it is important to be clear what outputs should take priority because the mechanisms for implementation will be different and in some cases contradictory, as was demonstrated in the case of the impact of different management strategies on men and women.

Future strategies

Government and donor support to JFM is now entering a new phase. It is beginning the difficult and slow task of facilitating the development of capacity at all levels to implement the programme. At this stage in its development, there are a number of strategies which could be followed; each has different implications for practice and potential outcomes.

Alternative 1 – 'Enrichment Planting'
Identify gaps in the current programme and focus attention on bringing these issues up to keep pace with the rest of the programme, e.g., attention to women and impacts on their livelihoods, NTFPs, training, silviculture, co-ordination between donors, institutionalising action research mechanisms, etc.

Alternative 2 – Expansion
Further institutionalise JFM in state and national policy. Increase the diversity of JFM models; move from degraded lands to lands with reasonable forest cover.

Alternative 3 – Redirection of Goals from Forestry to Natural Resource Management
Use JFM as a launching pad or vehicle for reaching other broader objectives in alleviating rural poverty. Rural people's well-being depends on multiple resources and is affected by broad sets of social objectives. For example – other common pool resources,

integrated resource management, government – NGO collaboration, inter-village issues, local governance, rights and resources issues.

Perhaps it is not just a question of options but rather a need to integrate all three of these strategies in one process. They represent different parts of the same strategy and the logical progression from a single-sector entry point to a multiple-sector solution.

The last strategy requires new institutional frameworks and policies and would lead to a new promotional phase and programme cycle.

8.4 Towards a 'New Generation' of Understanding in Forestry

Perhaps one of the most important conclusions to emerge from an assessment of the development of participatory forestry is that it has become a significant part of forestry policy and practice region-wide and indeed worldwide.

Over the last 20 years in Nepal and India there have been significant changes in land tenure and rights; the development of improved resource management practices; the forging of new collaborative relationships between government, non-governmental and academic institutions; the development of participatory tools and the incorporation of social science methods into the practices of the forestry sector; and lastly the beginnings of the institutionalisation process of these approaches. These changes in government attitudes and practice should enable participatory forestry programmes to have a country- or State-wide impact, to be sustainable because of the stability of resources available to the government, and to evolve to meet villagers' continually changing needs through an accountable decision-making process. In adopting these measures in which local people play a more dominant role, it can be said that the agencies have taken on a broader social and economic development agenda and contributed to fundamental shifts in the relationship

The new generation: joint management of forests to ensure future sustainability

between government and the rural civil society.

The experiences illustrated in this Guide do begin to chart the outlines of a fundamentally different form of forestry practice. It is now clear that there are at least four generalisable means of creating incentives for villagers to invest in forest and land conservation. These are:

- increasing villagers' participation in decision-making about forest land management
- increasing villagers' land-tenure security to encourage long-term investments in the resource base
- arranging the sharing of rights and responsibilities between the community and government to create incentives for joint management
- providing attractive alternative income opportunities that lessen the pressure for use of the forest.

In addition, lessons have been learned about how to work with large bureaucracies and how to support development generally. Much of the work carried out in participatory forestry programmes has had implications broader than those related to Forest Departments and people living in forest areas.

To conclude with a paraphrase of Westoby (1987), forestry is about trees and people, and their association through institutional arrangements. Wise forest management is about the development of careful partnerships, and conflict-resolution mechanisms and reconciling often divergent objectives. Forests are international, regional, national and local assets; as such the diversity of interest groups, objectives and management strategies is vast and complicated. Today's foresters confront a challenge ahead that compels them to face this complexity and diversity and respond with site-specific approaches, within strategic frameworks, that recognise the web of interrelated relationships of which they are a part. It is not so much a question of a new forestry but rather of a forestry that takes full cognisance of the social, economic, political, ecological and cultural contexts in which it operates.

In the strict sense of the understanding of paradigm (Kuhn, 1970) it is not possible for a new paradigm to emerge until a new generation of proponents also arises:

> A new scientific truth does not triumph by convincing its opponents and making them see the light, but rather because its opponents eventually die, and a new generation grows up familiar with it. (Kuhn, 1970)

Thus participatory forestry lies at the edge of change where contradictions, obvious within the working of the old practices, have not been overthrown by a new participatory or indeed 'pragmatic' forestry practice, but where a new generation is emerging which is demanding new partnerships for forest management. The future we are contemplating is one where there is no need for forestry to have qualifiers attached to it to explain the type of forestry in which one is engaged, rather all forms of tree management, under whatever form of institutional arrangement, will be known as forestry. The plea for multi-resource management made by Behan (1990) is one that fits with the approaches being tried in both India and Nepal. Forests and forestry are a complex series of interrelated resources and systems, a concept well explained by Henry H. Carey (1989):

> People often ask me: 'Why aren't our forests sustainable? Aren't we replanting after we cut?' My response is that a forest is not a plantation. A forest is a mosaic, including ancient trees, venerable giants providing shelter for wildlife. A forest is clear, pure water, rivers alive with trout and salmon. A forest is wildness, silence. A forest is tall, straight trees for house beams and logs. A forest is a wellspring of value for rural people. A forest is the deep, rich soil, soaking and storing water and nutrients.
>
> The fact that we can grow young trees where all this once stood does not mean we have perpetuated the forest. The task of protecting and restoring the forest is more complex. It demands all the skills, imagination and dedication of foresters. This is the art of forestry.

We cannot intervene in one part of the system without understanding the impact on the other interdependent parts. Hence, changes in silvicultural practice should be considered in the light of their impact on local livelihoods, biodiversity, market values and the consequences for future generations.

The final word therefore for a book that started out with a discussion of the evolution of participatory forestry as a policy and practice must be a return to the beginning to say that this is a period of change and a time of critical reflection (Behan, 1990; Umans, 1993). However, as with all discourse, the reality of endeavour and the lived experiences of individual forest users are those that should remain with us in our attempts to develop jointly systems of management that will help to secure future livelihoods instead of jeopardising them:

> For the poor people I do not think there will be an equal distribution of firewood ... The rich people have their own trees, they can cut them down for firewood. Poor people like us need to collect firewood these days because we have to go for wage labour during the summer. If we don't work for wages we will have to go hungry ... They said that poor people will have to cut branches and firewood which will be sold by weight ... I don't understand what they are doing and what they are talking about ... Rich and poor will be treated equally, they say, but they have

> asked the poor to dig the paths and plant the seedlings. If we work for wages in agriculture ... how can we dig paths without getting any money, what do we eat? Can the poor work with empty stomachs? (Thulo Kami, quoted in Hobley, 1990)

Perhaps, then, we are moving back to the 1960s and the words of Westoby (1962) when he urged forestry to move into the industrial age. Is this going to be the next movement – forestry for small-scale industrial development: from subsistence to the market? Or is it an additional charge to be added to the objectives of participatory forestry? Perhaps instead of looking for new words and objectives to describe our actions we should be looking for actions and change that support the livelihoods of people. We should move towards a new pragmatism in which we accept the boundaries of the possible for forestry and do not impose objectives which cannot be achieved by one sector alone.

Whatever the future may hold, we urge a move away from rhetoric to action in which the words we use address the felt needs of people. In the words of Humpty Dumpty and Alice:

> 'When I use a word', Humpty Dumpty said in a rather scornful voice, 'it means what I choose it to mean – neither more nor less.'
>
> 'The question is', said Alice, 'whether you can make words mean so many different things.'
>
> 'The question is', said Humpty Dumpty, 'which is to be the master – that's all?'

Bibliography

Acharya, H. (1984) 'Management of forest resources in Nepal: a case study of Madan Pokhara village panchayat'. Unpublished Master thesis, Cornell University, Ithaca, N.Y.

Acharya, H. (1989) 'Jirel property arrangements and the management of forest and pasture resources in highland Nepal', *Development Anthropology Network*. Binghamton, N.Y.: Institute for Development Anthropology.

Adhikari, J. (1990) 'Is community forestry a new concept? An analysis of the past and present policies affecting forest management in Nepal', *Society and Natural Resources* 3(3):257–65.

Adnan, S., Barrett, A., Nurul Alam, S.M. and Brustinow, A. (1992) *People's Participation. NGOs and the Flood Action Plan*, Dhaka: Research and Advisory Services.

Agarwal, A. and Narain, S. (1989) *Towards Green Villages*. Delhi: Centre for Science and Environment.

Agarwal, B. (1992) 'The gender and environment debate: lessons from India', *Feminist Studies* 18(1): 119–157.

Aggarwal, K.L. (1949) 'Fourth working plan for the Kulu-Seraj forests', Shimla: Himachal Pradesh Forest Department.

Agrawal, A. (1994) 'Rules, rule making, and rule breaking: examining the fit between rule systems and resource use'. In E.Ostrom, R.Gardner and J.Walker (eds) *Rules, Games and Common-Pool Resources*. Ann Arbor, MI: University of Michigan Press.

Alvares, C. (1982) 'A new mystification? Some social forestry is old practice rhetorically dressed up', *Development Forum* Jan-Feb:3–4.

Amery, C.F. (1876) 'On forest rights in India'. In D Brandis and A. Smythies (eds) 'Report on the Proceedings of the Forest Conference, Simla October 1875'. Calcutta: Government of India.

Anderson, A. (1886) 'Report on the Demarcation and Settlement of Kulu Forests', reprinted in 1975. Shimla : Himachal Pradesh Forest Department.

Anderson, R.S. and Huber W. (1988) *The Hour of the Fox: Tropical Forests, the World Bank and Indigenous People in Central India*. New Delhi: Vistaar/Sage.

Anon. (1885) 'Forest Conservancy in Bombay', *The Indian Forester*. XI(7):299–301.

Anon. (1922) 'Report of the inaugural meeting of the Empire Forestry Association', *Empire Forestry* 1: 3–10.

Anon. (1981) 'Statism not the answer', *Economic and Political Weekly* XVI(36): 1447–8.

Arnold, J.E.M. (1989) 'People's participation in forest and tree resource management: a review of ten years of community forestry', Rome: FAO.

Arnold, J.E.M. (1990) *Social Forestry and Communal Management in India*. Social Forestry Network Paper No. 11b, London: ODI.

Arnold, J.E.M. (1995) 'Retrospect and prospect'. In J.E.M. Arnold and P.A. Dewees (eds) *Tree Management in Farmer Strategies: Responses to Agricultural Intensification*. Oxford: Oxford University Press.

Arnold, J.E.M. and Bergman, A. (1988) *Forestry for the Poor? An Evaluation of the SIDA Supported Social Forestry Project*. New Delhi: SIDA.

Arnold, J.E.M. and Campbell, J.G. (1986) 'Collective management of hill forests in Nepal: the Community Forestry Development Project'. In *Proceedings of the Conference on Common Property Resource Management*, April 21–26, 1985. Washington, DC: National Academy Press.

Arnold, J.E.M. and Dewees, P.A. (eds) (1995) *Tree Management in Farmer Strategies: Responses to Agricultural Intensification*, Oxford: Oxford University Press.

Arnold, J.E.M. and Stewart, W.C. (1991) *Common Property Resource Management in India* . Tropical Forestry Papers No.24, Oxford: Oxford Forestry Institute, University of Oxford.

Arnold, J.E.M., Bergman, A. Harris, P. Mohanty, J. (1987a) 'Evaluation of the SIDA supported social forestry project in Orissa, India'. Report for SIDA, Stockholm.

Arnold, J.E.M., Bergman, A., Djurfeldt, G. (1987b) 'Evaluation of the SIDA supported social forestry project in Tamil Nadu, India'. Report for SIDA, Stockholm.

Arora, D. (1994) 'From state regulations to people's participation: case of forest management in India', *Economic and Political Weekly* XXIX(12).

Arora, H. and Khare, A. (1994) 'Experience with the recent joint forest management approach'. Paper prepared for the International Workshop on India's Forest Management and Ecological Revival, New Delhi 10–12 February. University of Florida and Tata Energy Research Institute, New Delhi.

Arora, S.S., Vijh, R. and Varalakshmi, V. (1993) *Yield of Bhabbar and Grasses from the Areas Managed by HRMS and Ballarpur Paper Mill – a Comparative Analysis*. Joint Forest Management Series No. 5, New Delhi: Haryana Forest Department and Tata Energy Research Institute.

Aryal, M. (1994) 'Axing Chipko', *Himal* Jan/Feb: 8–23.

Baden-Powell, B.H. (1874) *Land Systems of British India*. Vol. 1, Oxford: Clarendon Press.

Bahaguna, S. (1984) 'What price this hill development?', *Yojana* 28(8).

Bahaguna, V.K. (1992) *Collective Resource Management: An Experience in Harda Forest Division*. Bhopal: Regional Centre for Wastelands Development, Indian Institute of Forest Management.

Bahaguna, V.K., Luthra, V., and Rathore, B.M.S. (1994) 'Collective forest management in India', *Ambio* 23(4–5):269–73.

Baidya, K. (1985) 'Policy issues: the unresolved question of Indian social forestry – a viewpoint', *International Journal of Environmental Studies* 24:205–16.

Bajaj, J.L. and Sharma, R. (1995) 'Improving government delivery systems: some issues and prospects', *Economic and Political Weekly* 27 May, M73–80.

Bajracharya, D. (1983a) 'Deforestation in the food/fuel context historical and political perspectives from Nepal', *Mountain Research and Development* 3(3):227–40.

Bajracharya, D. (1983b) 'Fuel, food or forests? Dilemmas in a Nepali village', *World Development* 11(12): 1057–74.

Balooni, K. and Singh, K. (1993) 'Role of NGOs in afforestation of village common lands: a case study in a tribal area in Gujarat, India', Anand, Gujarat: Institute of Rural Management, (mimeo).

Baral, J.C. (1991) 'Indigenous forestry activities in Accham District of the Far Western Hills of Nepal', Field Document No. 15, HMGN/UNDP/FAO/CFDP, Kathamandu.

Baral, J.C. (1993) 'User group participation in community plantation: the approach and experience of Palpa district', *Banko Janakari* 4(1):67–73.

Baral, J.C. and Lamsal, P. (1991) 'Indigenous systems of forest management: a potential asset for implementing a community forestry programme in the hills of Nepal (with special reference to Palpa District)'. Field Document No. 17. Kathamandu: HMGN/UNDP/FAO/CFDP.

Baral, N.R. (1993) 'Where is our community forestry?, *Banko Janakari* 4(1)12–15.

Bartlett, A.G. (1992) 'A review of community forestry advances in Nepal', *Commonwealth Forestry Review* 71(2):95–100.

Bartlett, A.G. and Malla, Y.B. (1992) 'Local forest management and forest policy in Nepal', *Journal of World Forest Resource Management* 6: 99–116.

Bartlett, A.G., Nurse, M.C., Chhetri, R.B., and Kharel, S. (1992) 'Towards effective community forestry through forest user groups'. Kathmandu: NAFP Discussion Paper Series.

Bass, S., Dalal-Clayton, B., and Pretty, J. (1995) *Participation in Strategies for Sustainable Development'* Environmental Planning Issues No.7. London: IIED.

Behan, R.W. (1990) 'Multiresource forest management: a paradigmatic challenge to professional forestry', *Journal of Forestry*:12–18.

Beotra, B.R. (1926) *Preliminary and Settlement Reports on the Forests in Suket State* Sundernagar, India.

Bhati, J.P. (1990) *Development Strategies in Himachal Pradesh.* Mountain Farming Series No. 6. Kathmandu: ICIMOD.

Bhattacharya, P. (1994) 'Sal forest ecosystem: people need vis a vis forest management practices in West Bengal, India'. Paper presented at International Conference on Participatory Forest Management: Enabling Environment, Institute of Bio-social Research and Development and the Rural Development Centre, Indian Institute of Technology – Kharagpur, Calcutta, 5–7 December.

Bilham, R. (1994) 'The next great earthquake', *Himal* May/June.

Bista, D.B. (1987) *People of Nepal.* Kathmandu: Ratna Pustak Bhandar.

Blaikie, P. (1990) 'The state of land management policy – present and future'. Paper presented at the Rural Development Studies Seminar at the Institute of Social Studies, The Hague, 4 November.

Borrini, G. (1992) *Environment and Health as a Sustainable State.* Rome: International Course for Primary Health Care Managers (ICHM).

Branney, P. (1994) 'Guidelines for managing community forests in the hills of Nepal'. Kathmandu: NUKCFP Report No. E/05.

Branney, P. and Dev, O.P. (1993) 'Development of participatory forest management for community forests in the Koshi Hills', *Banko Janakari* 4(1):47–52.

Branney, P. and Dev, O.P. (1994a) 'Guidelines for preparing working plans for community forests in the hills of Nepal, (2nd edition)' Kathmandu: NUKCFP Report No. E/06.

Branney, P. and Dev, O.P. (1994b) 'Forest product yields from community forests: Preliminary data from the Koshi Hills'. Kathmandu: NUKCFP.

Britt-Kapoor, C. (1994) *A Tale of Two Committees: Villager Perspectives on Local Institutions, Forest Management and Resource Use in Two Central Himalayan Indian Villages* . Rural Development Forestry Network Paper 17a, London: ODI.

Bromley, D. (ed.) (1992) *Making the Commons Work: Theory, Practice and Policy.* San Francisco, CA: ICS Press.

Brown, C.L. and Valentine, J. (1994) 'The process and implications of privatisation for forestry institutions: focus on New Zealand', *Unasylva* 45(178):11–19.

Buchy, M. in press *Teak and Arecanut. Colonial State, Forest and People in the Western Ghats (South India), 1800–1947.* Pondicherry: IFPI/IGNA.

Budhathoki, P. (1987) 'Importance of community forestry management in remote areas: experience in Jajarkot District', *Banko Janakari* 1(4):24–29.

Byron, R.N. and Ohlsson, B. (1989) *Learning from the Farmers about their Trees*. FAO/SIDA Forest, Trees and People Programme. Rome: FAO.

Campbell J.G., Shrestha, R.J. and Euphart, F. (1987) 'Socioeconomic factors in traditional forest use and management: preliminary results from a study of community forest management in Nepal', *Banko Janakari* 1(4):45–54

Campbell, J.G and Denholm, J. (eds) (1993) *Inspirations in Community Forestry*. Report of the Seminar on Himalayan Community Forestry, Kathmandu, Nepal 1–4 June, Kathmandu: ICIMOD.

Campbell, J.G. (1978) 'Community involvement in conservation: social and organisational aspects of the proposed Resource Conservation and Utilisation Project in Nepal'. Report to the USAID, Nepal.

Campbell, Jeffrey Y. (1993) Changing objectives, new products, and management challenges, making the shift from 'major' vs. 'minor' forest products. Paper presented at the Seminar on Forest Produce, ICFRE, October 1993, Coimbatore.

Campbell, Jeffrey Y. (1995) 'Evolving forest maangement systems to people's needs'. In S.B. Roy (ed.) *Enabling Environment for Joint Forest Management*. New Delhi: Inter India Publications.

Campbell, Jeffrey Y., Palit, S., and Roy, S.B. (1994) 'Putting research partnerships to work: the joint forest management research network in India'. Paper prepared for the Fifth International Symposium on Society and Resource Management, 7–10 June, , Fort Collins, CO.

Carew-Reid, J., Prescott-Allen, R., Bass, S. and Dalal-Clayton, B. (1994) *Strategies for National Sustainable Development: A Handbook for their Planning and Implementation*. London and Geneva: IIED and IUCN.

Carey, H.H. (1989) 'Forest Trust: two-year report 1987–88.' Santa Fe, CA: The Forest Trust.

Carroll, T. (1992) *Intermediary NGOs: The Supporting Link in Grassroots Development*: West Hartford,CT: Kumarian Press.

Carter, E.J. (1992) *Tree Cultivation on Private Land in Nepal's Middle Hills: an Investigation into Local Knowledge and Local Needs*. OFI Occasional Paper 40 Oxford: Oxford Forestry Institute.

Carter, E.J. (1996) *Recent Approaches to Participatory Forest Resource Assessment*. Study Guide 2. London: ODI.

Carter, A.S. and Gilmour, D.A. (1989) 'Tree cover increases on private farm land in central Nepal', *Mountain Research and Development* 9(4):381–391

Cernea, M.M. (1992) 'A sociological framework: policy, environment, and the social actors for tree planting'. In N.P. Sharma (ed.) *Managing the World's Forests. Looking for Balance Between Conservation and Development*. Iowa: Kendall/Hunt Publishing Company.

Chaffey, D., Aziz, A., Djurfeldt, G., Haldin, G., Kohlin, G., Paranjape, J., Shyamsunder, S.,Svanquist, N. and Tejwani, K.J. (1992) 'An evaluation of the SIDA supported social forestry projects in Tamil Nadu and Orissa, India'. Report to SIDA, Stockholm.

Chambers, R. (1993) *Challenging the Professions: Frontiers for Rural Development*. London: Intermediate Technology Development Group.

Chambers, R., Saxena, N.C. and Shah, T. (1989) *To the Hands of the Poor: Water and Trees*. London: Intermediate Technology Publications.

Chatterjee, M. (1994) 'Gender roles in joint forest management', *Wasteland News* IX(4):23–5.

Chatterjee, M. (n.d.) *Women in Joint Forest Management: a Case Study from West Bengal*. IBRAD Technical Paper No. 4. Calcutta: IBRAD.

Chatterjee, M. and Roy, S.B. (1994) *Reflections from Training on Gender Issues in Joint Forest Management*. IBRAD: Calcutta.

Chatterji, J. and Gulati, M. (n.d.) 'Co-managing the commons: the J & K experience'. New Delhi: Society for Promotion of Wastelands Development.

Chaturvedi, A.N. (1992a) 'Management of secondary forests', *Wasteland News* 7(1).

Chaturvedi, A.N. (1992b) 'No, it is non-viable and unsound: a rejoinder', *Wasteland News* 7(1).

Chhetri, R.B. and Nurse, M.C. (1992) 'Equity in user group forestry: implementation of community forestry in Central Nepal'. *Discussion Paper Series*, Kathmandu: NACFP.

Chhetri, R.B. and Pandey, T.R. (1992) *User Group Forestry in the Far-Western Region of Nepal (case studies from Baitadi and Achham,* Kathmandu: ICIMOD.

Chhetri, R.B., Nurse, M.C. and Baral, S.P. (1993) 'Self-reliance among user groups: towards sustainable community forestry in Nepal', *Banko Janakari* 4(1):53–6.

Chopra, K. (1995) 'Forest and other sectors: critical role of government policy', *Economic and Political Weekly* 24 June:1480–2.

Chowdhury, K. (1992) 'Structural changes for better', *The Hindu Survey of the Environment*:31–3.

Ciriacy-Wantrup, S.V. and Bishop, R.C. (1975) 'Common property as a concept in natural resources policy', *Natural Resources Journal* 15:713–27.

Colchester, M. (1989) *Pirates, Squatters and Poachers: The Political Economy of Dispossession of the Native Peoples of Sarawak*. Kuala Lumpur: Survival International and INSAN.

Colchester, M. (1994) 'Sustaining the forests: the community-based approach in South and South-East Asia', *Development and Change* 25(1):69–100.

Collier, J.V. (1976) (1928) 'Forestry in Nepal'. Appendix 19 in Perceval Landon, *Nepal* republished 1976. Kathmandu: Ratna Pustak Bhandar.

Commander, S. (1986) *Managing Indian Forests: a Case for the Reform of Property Rights*. Social Forestry Network Paper No. 3b. London: ODI.

Cronin, E.W. (1979) *The Arun, a Natural History of the World's Deepest Valley*. Boston, MA: Houghton Mifflin.

CSE (1982) *The State of India's Environment, 1982: A Citizen's Report*. New Delhi: Centre for Science and Environment.

D'Abreo, D. (1982) *People and Forests: The Forest Bill and a New Forest Policy*. New Delhi: Indian Social Institute.

Dahal, D.R. (1994) *A Review of Forest User Groups: Case Studies from Eastern Nepal*. Kathmandu: ICIMOD.

Dahlman, C.J. (1980) *The Open Field System and Beyond*. Cambridge: Cambridge University Press.

Dani, A., Gibbs, C. and Bromley, D. (1987) *Institutional Development for Local Management of Rural Resources*. East-West Environment and Policy Institute Workshop report No. 2. Honolulu, Hawaii: East-West Center.

Dargavel, J., Hobley, M. and Kengen, S. (1985) 'Forestry of development and underdevelopment of forestry'. In J. Dargavel and G. Simpson (eds.) *Forestry: Success or Failure in Developing Countries*? CRES Working Paper 1985/20. Canberra: Centre for Resource and Environmental Studies, Australian National University.

Dasgupta, S.L. (1995) 'Forest bill: for whose benefit?,' *DEBACLE*. IX(1): 3–5.

Datta, S. (1995) *Joint Forest Management as a Process for Evolving an Appropriate Institutional System for Management of Degraded Forests in India – Lessons from Some Case Studies of West Bengal and Gujarat*. Ahmedabad: Report to Ford Foundation, Centre for Management in Agriculture, Indian Institute of Management.

Demsetz, H. (1967) 'Toward a theory of property rights', *American Economics Review* 62(2):347–59.

Dev, O.P. (1994) *A Manual for Community Forest Management Workshop*. Kathmandu: NUKCFP Report No. D1/01.

Dewees, P.A. (1995) 'Farmer responses to tree scarcity: the case of woodfuel'. In J.E.M. Arnold and P.A. Dewees (eds) *Tree Management in Farmer Strategies: Responses to Agricultural Intensification*. Oxford: Oxford University Press.

Dewees, P.A. and Saxena, N.C. (1995) 'Wood product markets as incentives for farmer tree growing'. In J.E.M. Arnold and P.A. Dewees (eds) *Tree Management in Farmer Strategies: Responses to Agricultural Intensification*. Oxford: Oxford University Press.

Dhar, S.K., Gupta, J.R. and Sarin, M. (1991). *Participatory Forest Management in the Shivalik Hills: Experiences of the Haryana Forest Department*. Sustainable Forest Management Series No. 5. New Delhi: Ford Foundation.

DN (1990) 'Women and forests', *Economic and Political Weekly* 14 April, 795–6.

Dogra, B. (1985) 'The World Bank vs the people of Bastar', *The Ecologist*. 15(1/2): 44–9.

Donovan, D. (1981) 'Fuelwood: how much do we need?', *Institute of Current World Affairs Newsletter* DGD 14, Hanover.

Douglas, J.J. (1983) *A Re-appraisal of Forestry Development in Developing Countries*. The Hague: Martinus Nijhoff/Dr W. Junk Publishers.

Down to Earth (1995) 'Whose forest is it anyway?', *Down to Earth* January.

Drona, K.C. (1994) 'Forest user group formation process and its impact: case studies from Yangpang, Bhojpur'. Project Report No. 12. Kathmandu: NUKCFP.

Dutta, M. and Adhikari, M. (1991) *Sal Leaf Plate Making in West Bengal: a Case Study of the Cottage Industry in Sabalmara, West Midnapore*, Indian Institute of Bio-Social Research Working Paper No. 2. Calcutta: IBRAD.

Eagle, S. (1992) Experiences with a Heterogeneous Forest User Group in the Far West of Nepal. Rural Development Forestry Network From the Field Summer No.13e, London: ODI.

Earl, D.E. (1975) *Forest Energy and Economic Development*, Oxford: Clarendon Press.

Eckholm, E.P. (1975) *The Other Energy Crisis: Firewood* Worldwatch Paper 1. Washington DC: Worldwatch Institute.

Eckholm, E.P. (1976) *Losing Ground: Environmental Stress and World Food Prospects*, New York: W.W Norton and Co.

Eckholm, E.P. (1979) *Planting for the Future: Forestry for Human Needs*, Worldwatch Paper No. 26. Washington DC: Worldwatch Institute.

Edwards, D. (1995) *Non-timber Forest Products from Nepal: The Medicinal and Aromatic Plant Trade'*. Occasional Paper. Kathmandu: Forest Research and Survey Centre.

Eisenstadt, E.N. (1966) *Modernisation: Protest and Change*. Englewood Cliffs, NJ: Prentice-Hall.

Elwin, V. (1964) *The Tribal World of Verrier Elwin: An Autobiography*. London: Oxford University Press.

Emerson, H.W. (1917) 'Note on the Report of the Forest Officer on the Mandi Forest Settlement', in *Report on the Mandi Forest Settlement*, Mandi State.

English, R. (1982) 'Gorkhali and Kiranti: political economy in the eastern hills of Nepal'. Ann Arbor, MI: University Microfilms International.

English, R. (1985) 'Himalayan state formation and the impact of British rule in the nineteenth century', *Mountain Research and Development* 5(1):65–76.

Eyben, R. (1991) 'The process approach [to project appraisal]'. In ODA, *Getting the Balance Right: How Should ODA Deal with Cross-disciplinary Issues*. Report of the Natural Resource Advisers Conference, University of Wales, Bangor, 8–11 July,. London: ODA.

FAO (1978) *Forestry for Local Community Development*. FAO Forestry Paper 7. Rome: FAO.

FAO (1980) 'Report of the FAO/SIDA seminar on forestry in rural community development. Rome: FAO.

Farrington, J., Bebbington, A. with Wellard, K. and Lewis, D.J. (1993) *Reluctant Partners: Non-Governmental Organisations, The State and Sustainable Agricultural Development*. 4 vols. London and New York: Routledge.

Femconsult (1995a) 'Study of the incentives for joint forest management'. Main Report prepared for the World Bank, Washington DC.

Femconsult (1995b) 'Incentives to joint forest management study. West Bengal field study – Bankura report'. Report to the World Bank, Washington DC.

Femconsult (1995c) 'Incentives to joint forest management study. Gujarat field study – Rajpipla report'. Report to the World Bank, Washington DC.

Femconsult (1995d) 'Historical background to joint forest management'. Working Paper No. 1. Report to the World Bank, Washington DC.

Fernandes, W. and Kulkarni, S. (1983) *Towards a New Forest Policy: People's Rights and Environmental Needs*. New Delhi: Indian Social Institute.

Fernandes, W. and Menon, G. (1987) *Tribal Women and Forest Economy*. Indian Social Institute, Tribes of India Series 1. New Delhi: ISI.

Fernandes, W. and Tandon, R. (1981) *Participatory Research and Evaluation: Experiments in Research as a Process of Liberation*. New Delhi: Indian Social Institute.

Fernandes, W., Menon, G. and Viegas, P. (1984) *Forests, Environment and Forest Dweller Economy in Orissa*. New Delhi: Indian Social Institute.

Fernandes, W., Menon, G. and Viegas, P. (1988) *Forests, Environment and Tribal Economy*. Indian Social Institute, Tribes of India Series No. 2. New Delhi: Indian Social Institute.

Fisher, R.J. (1989) *Indigenous Systems of Common Property Forest Management in Nepal* Working Paper No. 18. Honolulu, Hawaii: Environment and Policy Institute, East-West Center.

Fisher, R.J. (1991) *Studying Indigenous Systems of Common Property Forest Management in Nepal: Towards a More Systematic Approach*. Working Paper No. 30. Honolulu, Hawaii: East-West Center.

Fisher, R.J. (1994) 'Indigenous forest management in Nepal: why common property is not a problem'. In M. Allen (ed) *Anthropology of Nepal: People, Problem and Processes*. Kathmandu: Mandela Book Point.

Fisher, R.J. (1995) *Collaborative Management of Forests for Conservation and Development*. IUCN/WWF Issues in Forest Conservation Series. Gland, Switzerland: IUCN.

Fisher, R.J., Singh, H.K., Pandey, D.R. and Lang, H. (1989) *The Management of Forest Resources in Rural Development: a Case Study of Sindhu Palchok and Kabhre Palanchok Districts of Nepal*. Mountain Populations and Institutions Discussion Paper No.1. Kathmandu: ICIMOD.

Foley, G. and Barnard, G. (1984) *Farm and Community Forestry*. Technical report No. 3. London: Earthscan.

Fortmann, L. and Bruce, D. (1988) *Whose Trees? Proprietary Dimensions of Forestry*. Boulder, CO: Westview Press.

Fox, J.M. (1991) 'Spatial Information for Resource Management in Asia: A Review of Institutional Issues', *International Journal of Geographical Information Systems* 5(1):59–72.

Fox, J.M. (1992) *The Problem of Scale in Community Resource Management*. East-West Center Reprints, Environment Series No. 5. Honolulu, Hawaii: East-West Center.

Fox, J.M. (1993) 'Forest resources in a Nepali village in 1980 and 1990: the positive influence of population growth', *Mountain Research and Development* 13(1):89–98.

Frank, A.G. (1969) 'Sociology of development and underdevelopment of sociology'. In A.G. Frank *Latin America: Underdevelopment of Revolution*. New York: Monthly Press.

FRI (1961) *100 Years of Forestry in India*. Dehra Dun: FRI.

Furer-Haimendorf, C. von (1964) *Sherpas of Nepal: Buddhist Highlanders*. London: John Murray.

Gadgil M. and Guha, R. (1992a) *This Fissured Land: An Ecological History of India*. New Delhi: Oxford University Press.

Gadgil, M. and Guha, R. (1992b) 'For genuine friendship', *The Hindu Survey of the Environment*, 26–9.

Gadgil, M. and Guha, R. (1992c) 'Interpreting Indian environmentalism'. Paper presented at Conference on the Social Dimensions of Environment and Sustainable Development Valletta, Malta, 22–25 April.

Gadgil, M., Prasad, S.N. and Ali, R. (1983) 'Forest management and forest policy in India: a critical review', *Social Action*. 33: 127–55.

Gautam, K.H. (1992) 'Indigenous forest management systems in the hills of Nepal'. Unpublished MSc. thesis, Australian National University, Canberra.

Gayfer, J.R. (1994) Nepal-UK Community Forestry Project: An Overview, April 1994. Kathmandu: NUKCFP.

Gayfer, J.R. and Pokharel, B.K. (1993) 'Bottom-up planning: an example of district annual plan and budget preparation from the Koshi Hills', *Banko Janakari* 4(1):85–90.

Ghai, D. (1994) 'Environment, livelihood and empowerment' *Development and Change* 25(1):1–11.

Ghai, D. and Vivian, J.(eds) (1992) *Grassroots Environmental Action: People's Participation in Sustainable Development*. London and New York: Routledge.

Ghai, D., Khan, A.R., Lee, E.L.H. and Alfthan, T. (1979) *The Basic Needs Approach to Development*. Geneva: ILO.

Ghose, A. (1995) 'Joint Forest Management Programme'. Report for People and Participation Course, Centre for Rural Development Training, University of Wolverhampton (mimeo).

Ghosh, A. (1994) 'Ideologues and Ideology: Privatisation of Public Enterprises', *Economic and Political Weekly* xxix (30):1929–31.

Gibbs, C. (1986) 'Institutional and organisational concerns in upper watershed management. In K. Easter, J. Dixon and M. Hufschmidt (eds) *Watershed Resources Management: An Integrated Framework with Studies from Asia and the Pacific*. Boulder, CO: Westview.

Gilmour, D.A. (1990) 'Resource availability and indigenous forest management systems in Nepal', *Society and Natural Resources*. 3:145–58.

Gilmour, D.A and Fisher, R.J. (1991) *Villagers, Forests and Foresters*. Kathmandu: Sahayogi Press.

Gilmour, D.A. and Nurse, M. (1991) 'Farmer initiatives in increasing tree cover in Central Nepal. Paper presented at workshop on Socio-economic Aspects of Tree Growing by Farmers. Institute of Rural Management, Anand, Gujarat.

Gilmour, D.A., King, G.C. and Hobley, M. (1989) 'Management of forests for local use in the hills of Nepal. 1. Changing forest management paradigms', *Journal of World Forests Resource Management* 4(2): 93–110.

Glueck, P. (1986) 'Social values embodied in the forestry profession'. Paper submitted to 18th IUFRO World Congress, Ljublijana.

GOI (1952) 'The National Forest Policy of India'. New Delhi: Government of India.

GOI (1976) *Report of the National Commission on Agriculture: Forestry, Volume IX*. New Delhi: Ministry of Agriculture and Irrigation.

GOI (1990) *Developing India's Wastelands*. New Delhi: Ministry of Environment and Forests.

Gordon, W.A. (1955) *The Law of Forestry* London: HMSO.

Gore, Al (1993) *From Red Tape to Results: Creating a Government that Works Better and Costs Less*. New York: Times Books.

Gregersen, H., Arnold, J.E.M., Lundgren, A., Contreras, A., de Montalembert, M.R. and Gow, D. (1993) *Assessing Forestry Project Impacts: Issues and Strategies*. FAO Forestry Paper 114. Rome: FAO.

Gregory, G.R. (1965) 'Forests and economic development in Latin America', *Journal of Forestry* 63(2):83–8.

Griffin, D.M. (1987) 'Implementation failure caused by institutional problems', *Mountain Research and Development* 7(3):250–3.

Griffin, D.M. (1988) *Innocents Abroad in the Forests of Nepal: An Account of Australian Aid to Nepalese forestry.* Canberra: Anutech.

Griffin, K. and Khan, A.R. (1978) 'Poverty in the Third World: ugly facts and fancy models', *World Development* 6(3):295–304.

Grimble, R.G., Aglionby, J. and Quan, J. (1994) *Tree Resources and Environmental Policy: A Stakeholder Approach*. NRI Socio-economic Series 7. Chatham, UK: Natural Resources Institute.

Gronow, J. and Shrestha, N.K. (1990) *From Policing to Participation: Reorientation of Forest Department Field Staff in Nepal*. Winrock Research Report Series No. 11. Kathmandu: Winrock International.

Guha, R. (1983) 'Forestry in British and Post-British India: A Historical Analysis', *Economic and Political Weekly*, 29 October and 5–12 November.

Guha, R. (1989) *The Unquiet Woods: Ecological Change and Peasant Resistance in The Himalayas*. New Delhi: Oxford University Press and; Berkeley, CA: University of California Press.

Guha, R. (1994) 'Forestry debate and draft forest act: who wins who loses', *Economic and Political Weekly* xxix(3):2192–6.

Guhathakurta, P. (1984) 'Experience with India's social forestry', *Indian and Foreign Review* July .

Guhathakurta, P. (1992a) 'Is management of coppice sal forests on short rotations sustainable?, *Wasteland News* 7(1).

Guhathakurta, P. (1992b) 'Switch-over from uni-tier to multi-tier plantations', *Wasteland News* 7(1).

Gupta, A. (1994) 'Nepali Congress and post-panchayat politics', *Economic and Political Weekly*, October: 2798–801.

Hamilton, C. (1985) 'The returns of social forestry: a cost-benefit analysis of the Nepal-Australia Forestry Project phase 3'. Centre for Development Studies Working Paper No. 85. Canberra: Australian National University.

Hamilton, F. (1971) (1819) *An Account of the Kingdom of Nepal*. 1971 reprint of 1819 edition. New Delhi: Manjusri Publishing House.

Hardin, G. (1968) 'The Tragedy in the Commons', *Science* 162:1243–8.

Hardin, G. (1994) 'The tragedy of the unmanaged commons', *Trends in Ecology and Evolution* 9.

Harriss, J. (ed) (1982) *Rural Development: Theories of Peasant Economy and Agrarian Change*. London: Hutchinson University Library.

Hausler, S. (1993) 'Community forestry: a critical assessment', *The Ecologist* 23(3):84–90

Higgott, R. (1978) 'Competing theoretical perspectives on development and underdevelopment: a recent intellectual history', *Politics* 23(1):26–41.

Hirsch, P. (1993) 'The State in the village: The case of Ban Mai', *The Ecologist*, 23(6):205–11.

HMGN (1990) 'Master Plan for the Forest Sector, Nepal, Forestry Sector Policy Revised Edition'. Kathmandu: Ministry of Forests and Environment/Finnida/Asian Development Bank.

HMGN (1992) *Operational Guidelines of the Community Forestry Programme*. Kathmandu: Community Forestry Development Division, Ministry of Forest and Environment.

Hobley, M. (1985) 'Common property does not cause deforestation', *Journal of Forestry* 83: 663–4.

Hobley, M. (1990) 'Social reality, social forestry: the case of two Nepalese panchayats'. PhD thesis, Australian National University, Canberra.

Hobley, M. (1991) 'Gender, Class and the Use of Forest Resources: Nepal'. In A. Rodda (ed.) *Women and the Environment,* London: Zed Press.

Hobley, M. (1992a) *Policy, Rights and Local Forest Management: The Case of Himachal Pradesh, India.* Rural Development Forestry Network Paper 13b. London: ODI.

Hobley, M. (1992b) 'The Terai Community Forestry Project'. Working paper prepared for the World Bank project preparation mission. Kathmandu: World Bank.

Hobley, M., Campbell, Jeffrey, Y. and Bhatia, A. (1994) 'Community forestry in India and Nepal: learning from each other'. In D.N. Tewari (ed.) *Community Forestry in India and Nepal: Sharing Experiences in Himalayan Ecosystems*. Dehra Dun: Indian Book Distributors.

Hodgson, (1972) (1841) *Essays on the Language, Literature and Religion of Nepal and Tibet.* Reprinted with additions in 1972. Amsterdam: Philo Press.

Hoffpauir, R. (1978) 'Deforestation in the Nepal Himalaya: a village perspective', *Association of Pacific Coast Geographers* 40:79–89.

IFS (1894) 'Appendix XVII Forest Policy'. In *Appendices to Forest Department Code. Circular No. 22F* 19 October, Government of India.

IFS (1894) *Review of Forest Administration in British India for 1892–1893*. Government of India.

Ingles, A. W. (1994) 'The influence of religious beliefs and rituals on forest conservation in Nepal'. Discussion Paper. Kathmandu: NACFP.

ITTO (1990) *Guidelines for the Sustainable Management of Natural Tropical Forests*. Yokohama: ITTO.

Ives, J.D. (1987) 'The theory of Himalayan environmental degradation: its validity and application challenged by recent research', *Mountain Research and Development* 7:189–199.

Ives, J.D. and Messerli, B. (1989) *The Himalayan Dilemma: Reconciling Development and Conservation*. London and New York: Routledge and the United Nations University.

Jackson, C. (1993) 'Doing what comes naturally? Women and environment in development', *World Development* 21(12):1947–63.

Jackson, W.J. (1989) 'Reorientation of forestry field staff'. Paper presented at the National Workshop on Planning for Community Forestry Projects. State Planning Institute and Forest Department of Government of Uttar Pradesh, Lucknow.

Jackson, W.J. (1990) 'Indigenous management of community forest resources in the middle hills of Nepal: a case study'. Unpublished Graduate Diploma in Science thesis, Australian National University, Canberra.

Jackson, W.J. (1993) 'Action research for community forestry: the case of the Nepal Australia Community Forestry Project'. Paper presented at ICIMOD Methodology Workshop on Rehabilitation of Degraded Mountatin Ecosystems of the Hindu-Kush Himalayan Region. 29 May-3 June, Kathmandu.

Jackson, W.J., Nurse, M.C., and Chhetri, R.B. (1993) 'High altitude forests in the Middle Hills of Nepal – can they be managed as community forest?, *Banko Janakari* 4(1):20–3.

Jodha, N.S. (1995) 'Common property resources and the dynamics of rural poverty in India's dry regions', *Unasylva* 46(180): 23–9.

Joekes, S., Heyzer, N., Oniang'o, R. and Salles, V. (1994) 'Gender, environment and population', *Development and Change* 25(1):137–65.

Johnson, S. (1995) 'Clarified JFM concepts'. In S.B. Roy (ed.) *Enabling Environment for Joint Forest Management*. New Delhi: Inter-India Publications.

Johri, B.M. and Babu, C.R. (1974) 'Forests enrich man's life in a hundred ways', *Yojana* XVIII (11):15–18.

Joshi, A.L. (1993) 'Effects on administration of changed forest policies in Nepal'. In *Policy and Legislation in Community Forestry*, Proceedings of a workshop 27–29 January. Bangkok: RECOFTC.

Joshi, B.L. and Rose L.E. (1966) *Democratic Innovations in Nepal: A Case Study of Political Acculturation*. Berkeley, CA: University of California Press.

Kafle, G. (n.d.) *A Process for User Group Formation in Nepal: Problems and Solutions in Bhakimle and Ahale Forests, Bhojpur District*. Project Working Paper No. 3 Kathmandu: NUKCFP.

Kafle, G. and Tumbahampe, N. (1993) Forest User Groups' Networking Workshop: A Manual for Facilitators. Kathmandu: NUKCFP.

Kant, S., Singh, N.M., and Singh, K.K. (1991) *Community Based Forest Management Systems: Case Studies from Orissa*. ISO/ Swedforest. Bhopal: Indian Institute of Forest Management and SIDA.

Karki, M., Karki, J.B.S. and Karki, N. (1994) *Sustainable Management of Common Forest Resources: An Evaluation of Selected Forest User Groups in Western Nepal*. Kathmandu: ICIMOD.

Kaul, O.N. (ed.) (1993) *Joint Forest Management in Haryana*. Joint Forest Management Series No. 15. New Delhi: Haryana Forest Department and Tata Energy Research Institute.

Kaul, O.N. and Dhar, S.K. (1994) 'Joint forest management in Haryana Shivaliks'. Paper prepared for International Workshop on India's Forest Management and Ecological Revival, New Delhi, 10–12 February, University of Florida and Tata Energy Research Institute, New Delhi.

Kayastha, B.P. (1990) 'Robin Hood cannot do it alone', *Himal* September/October.

Keay, R.W.G. (1971) 'The role of forests in the development of tropical countries'. Paper presented at the 12th Pacific Science Congress, August.

Keeling, S. (1993) 'How to form a community forestry user group and be given responsibility for management of your forest', *Woodpecker* (VSO forestry newsletter), June.

King G.C., Hobley, M. and Gilmour, D.A. (1990) 'Management of forests for local use in the hills of Nepal. Part II. Towards the development of participatory forest management.', *Journal of World Forest Resource Management* 5:1–13.

Kolavalli, S. (1995) 'Joint forest management: superior property rights?', *Economic and Political Weekly* XXX(30):1933–8.

Kondos, A. (1987) 'The question of 'corruption' in Nepal', *Mankind* 17(1): 15–29.

Korten, D. (1980) 'Community organisation and rural development: a learning process approach', *Public Adminstration Review*

KOSEVEG (1994) 'Management Skills Workshop and Development of Group Level Data Recording Forms', M. and E. Consultancy Report No. 2, July 1994. Kathmandu: ODA.

Kuhn, T.S. (1970) *The Structure of Scientific Revolutions*. Chicago, IL: University of Chicago Press.

Kulkarni, S. (1982) 'Encroachment on forests: government versus people', *Economic and Political Weekly* XVII (3):55–9.

Kulkarni, S. (1983) 'Towards a social forest policy', *Economic and Political Weekly* XVIII (6):191–6.

Lal, J.B. (1994) Silvicultural and institution revision for sustainable forest management. Abstract presented at the International Seminar on Management of NTFP, Centre for Minor Forest Products, 12–15 November, Dehra Dun.

Lerner, D. (1965) *The Passing of Traditional Societies: Modernising the Middle East*. New York: Free Press.

Leslie, A. (1985) 'A plague on all your houses: reflections on 20 years in the development of forestry for Third World development'. In J Dargavel and G. Simpson (eds) *Forestry: Success or Failure in Developing Countries*. CRES Working Paper. Canberra: Centre for Resource and Environmental Studies, Australian National University.

Leslie, A. (1987) 'Foreword'. In Westoby.

Leys, C. (1977) 'Underdevelopment and dependency: critical notes', *Journal of Contemporary Asia* 7(1):92–107.

Libecap, G.D. (1989) 'Distributional issues in contracting for property rights', *Journal of Institutional and Theoretical Economics* 145:6–24.

Lingam, L. (1994) 'Women-headed households: coping with caste, class and gender hierarchies', *Economic and Political Weekly* 19 March: 699–704.

Lohani, P.C. (1973) 'Industrial policy'. In P.S.J.B. Rana and K.P.Malla (eds) *Nepal in Perspective*. Kathmandu: Centre for Economic Development and Administration.

Loughhead, S., Shrestha, R. and Raj, D.R.C. (1994) 'Social development considerations in community forestry'. NUKCFP Working Paper Series, Project Working Paper No. 2. Kathmandu: NUKCFP.

Luthra, V. (1994) 'Lack of people's involvement in forest management: are foresters the villains', *Wasteland News* IX(3):45–9.

Lynch, O.J. and Talbott, K. (1995) *Balancing Acts: Community Based Forest Management and National Law in Asia and the Pacific.* Washington, DC: World Resources Institute.

Maharjan, M.R. (1994) 'Extent of illegal cutting: Is the protection role of the DFO still needed?' Kathmandu: NUKCFP, Draft Report No. B/13.

Mahat, T.B.S. (1985) 'Human impact on forests in the Middle Hills of Nepal'. Unpublished PhD thesis, Australian National University, Canberra.

Mahat, T.B.S., Griffin, D.M. and Shepherd, K.R. (1986) 'Human impact on some forests of the Middle Hills of Nepal. 1. Forestry in the context of the traditional resources of the state', *Mountain Research and Development* 6:223–232.

Mahat, T.B.S., Griffin, D.M. and Shepherd, K.R. (1987) 'Human impact on some forests of the Middle Hills of Nepal. 4. A detailed study in S.E. Sindhu Palchok and N.E. Kabhre Palanchok', *Mountain Research and Development* 7:111–34.

Mahiti Team (1983) 'Why is social forestry not 'social'? Paper presented at The Ford Foundation Workshop on Social Forestry and Voluntary Agencies, Badhkhal Lake, Haryana, 13–15 April.

Maithani, G.P. (1994) 'Management perspectives of minor forest products'. Paper presented at the International Seminar on Management of NTFP, Centre for Minor Forest Products, 12–15 November, Dehra Dun.

Malhotra, K.C. and Deb, D. (1991) 'History of deforestation and regeneration/plantation in Midnapore District of West Bengal, India'. Paper prepared for the IUFRO International Conference on History of Small Scale Private Forestry, 2–5 September, Freiburg.

Malhotra, K.C., Chandra Satish, N., Vasulu, T.S., Majumdar, L., Basu, S., Adhikari, M. and Yadav, G. (n.d.) *Joint Management of Forest Lands in West Bengal: a Case Study of Jamboni Range in Midnapore District*. Technical Paper No.2. Calcutta: IBRAD.

Malhotra, K.C., Deb, D., Dutta, M., Vasulu, T.S., Yadav, G. and Adhikari, M. (1991) *Role of Non-timber Forest Produce in Village Economy.* Calcutta: IBRAD.

Malhotra, K.C., Poffenberger, M., Bhattacharya, A. and Dev, D. (n.d.) *Rapid Rural Appraisal Methodology Trials in Southwest Bengal: Assessing Natural Forest Regeneration Patterns and Non-wood Forest Product Harvesting Practices*. Sustainable Forest Management Working Paper No. 11. New Delhi: Ford Foundation.

Malla, Y.B. (1992) 'The changing role of the forest resource in the hills of Nepal'. Unpublished PhD thesis, Australian National University, Canberra.

Malla, Y.B. (1993) 'Market: an ignored dimension of community forestry in Nepal', *Banko Janakari* 4(1):28–31.

Malla, Y.B., Jackson, W.J. and Ingles, A.W. (1989) *Community Forestry for Rural Development in Nepal. A Manual for Training Field Workers. Part I and Part II.* Kathmandu: Nepal Australia Forestry Project.

McGregor, J.J. (1976) 'The existing and potential roles of forestry in the economics of developing countries'. In *Evaluation of the Contribution of Forestry to Economic Development* Bulletin No. 56. Edinburgh: Forestry Commission.

McKean, M. (1986) 'Management of Traditional Common Lands in Japan'. In National Research Council, Proceedings of the Conference on Common Property Resource Management. Washington, DC: National Academy Press.

McKean, M. (1995) 'Common property: what is it, what is it good for, and what makes it work?' Paper presented at the International Conference on Chinese Rural Collectives and Voluntary Organisations: Between State Organisation and Private Interest, Sinological Institute, University of Leiden, 9–13 January.

McKean, M. and Ostrom, E. (1995) 'Common property regimes in the forest: just a relic from the past?', *Unasylva* 46(180):3–15.

Messerschmidt, D.A. (1981) 'Nogar and other traditional forms of cooperation in Nepal: significance for development, *Human Organisation* 40: 40–7.

Messerschmidt, D.A. (1984) *Using Human Resources in Natural Resource Management: Innovations in Himalayan Development*. Watershed Management Working Paper No. 1. Kathmandu: ICIMOD.

Messerschmidt, D.A. (1986) 'People and resources in Nepal: customary resource management systems of the upper Kali Gandaki'. In Proceedings of the Conference on Common Property Resource Management. Washington DC: National Academy Press.

Messerschmidt, D.A. (1987) 'Conservation and society in Nepal: traditional forest management and innovative development'. In P.Little and M. Horowitz (eds) *Lands at Risk in the Third World: Local Level Perspectives*. Boulder, CO: Westview Press.

Moench, M. (1990) *Training and Planning for Joint Forest Management*. Sustainable Forest Management Working Paper No. 8. New Delhi: Ford Foundation.

Molnar, A. (1981) 'The dynamics of traditional systems of forest management: implications for the Community Forestry Development and Training Project'. Consultant's Sociologist's Report to the World Bank, Washington, DC.

Molnar, A. and Schreiber, G. (1989) *Women and Forestry: Operational Issues* Washington, DC: World Bank.

Moris, J. and Copestake, J. (1993) *Qualitative Enquiry for Rural Development: A Review,* London: Overseas Development Institute/Intermediate Technology Publications.

Mosse, D. (1993) *Authority, Gender and Knowledge: Theoretical Reflections on the Practice of Participatory Rural Appraisal.* Agricultural Adminstration Research and Extension Network Paper 44, London: ODI.

Mukherjee, N. (1995) 'Forest management and survival needs', *Economic and Political Weekly* 9 December: 3130–2.

Mukul (1993) 'Villages of Chipko movement', *Economic and Political Weekly* 10 April: 617–20.

Muller-Boker, U. (1991) 'Knowledge and evaluation of the environment in traditional societies of Nepal', *Mountain Research and Development* 11(2):101–14.

Muranjan, S.W. (1974) 'Exploitation of forests through labour cooperatives', *Artha Vijnana* 16(2):101–226.

Muranjan, S.W. (1980) 'Impact of some policies of the forest development corporation on the working of the forest labourers' cooperatives', *Artha Vijnana* 22(4):485–511.

NACFP and Community and Private Forestry Division (1994) *A Brief Guide to Range Post Planning* Technical Note No. 6/94, Kathmandu: NACFP and CPFD.

Nadkarni, M.V. with Pasha, S.A. and Prabhakar, L.S. (1989) *The Political Economy of Forest Use and Management*. New Delhi: Sage Publications.

NAFP (1979) 'Nepal's National Forestry Plan 1976 (2033B.S.) (unofficial English translation)'. Kathmandu: Nepal-Australia Forestry Project.

NAFP (1985) *Nepal-Australia Forestry Project. Phase 3 Project Document*. Canberra: ANUTECH.

Narain, U. (1994) 'Women's involvement in joint forest management: analysing the issues'. Paper prepared for the Ford Foundation Joint Forest Management Working Paper Series, 6 May.

National Planning Commission, 1985 'Programmes for Fulfilment of Basic Needs'. Kathmandu: National Planning Commission.

Netting, R. McC. (1976) 'What alpine peasants have in common: observations on communal tenure in a Swiss village', *Human Ecology* 4:135–46.

Nield, R.S. (1985) 'Fuelwood and fodder – problems and policy'. Working Paper for the Water and Energy Commission Secretariat, Kathmandu.

Nurse, M.C. and Chhetri, R.B. (1992) 'Implementation of community forestry in Central Nepal: a methodology for monitoring and evaluation of operational plans and the district development programme'. Paper presented *at Seminar on Community Forestry in the Himalayan Region* 1–4 June, Kathmandu: ICIMOD.

ODA (1994) 'Stakeholder participation in aid activities'. ODA Technical Note, London: ODA.

ODA (1995) *A Guide to Social Analysis for Projects in Developing Countries*. London: HMSO.

Openshaw, K. (1974) 'Wood fuels the developing world', *New Scientist* 61:883.

Ostrom, E. (1990) *Governing the Commons: The Evolution of Institutions for Collective Action*, Cambridge: Cambridge University Press.

Ostrom, E. (1992) 'Community and the endogenous solution of commons problems', *Journal of Theoretical Politics* 4(3):343–51.

Ostrom, E. (1994) *Neither Market nor State: Governance of Common-Pool Resources in the Twenty-first Century.* IFPRI Lecture Series No. 2, Washington, DC: IFPRI.

Ostrom, V., Feeny, D., and Picht, H. (eds) (1988) *Rethinking Institutional Analysis and Development: Issues, Alternatives and Choices.* San Francisco, CA: International Center for Economic Growth.

Pachauri, R. (n.d.) *Sal Leaf Plate Processing and Marketing in West Bengal.* Sustainable Forest Management Working Paper No. 12, New Delhi: Ford Foundation.

Pal, M. (1994) 'Centralised Decentralisation: Haryana Panchayati Raj Act, 1994', *Economic and Political Weekly* xxix (29):1842–4.

Palit, S. (1991) 'Participatory management of forests in West Bengal, *The Indian Forester* 117(5):342–9.

Palit, S. (1993a) *The Future of Indian Forest Management: Into the Twenty-first Century.* Joint Forest Management Working Paper No. 15, New Delhi: Society for Promotion of Wastelands Development and The Ford Foundation.

Palit, S. (1993b) 'Implementation of joint forest management', *Wasteland News* VIII(2):58–66.

Panda, A., Dabas, M., Varalakshmi, V., Shah, V.N. and Gupta, J.R. (1992) *Grass Yield under Community Participation in Haryana Shivaliks.* Joint Forest Management Series No. 1. New Delhi: Haryana Forest Department and Tata Energy Research Institute.

Pandey, K.K. (1982) *Fodder Trees and Tree Fodder in Nepal.* Berne: Swiss Development Corporation.

Pandey, T.R. (1994) *Forest User Groups in Koshi Hills: a Note on Some Social Issues and Institutional Process.* Nepal-UK Community Forestry Project Report No. 3, Kathmandu: NUKCFP.

Pant, M.M. (1980) 'The impact of social forestry on the national economy of India', *The International Tree Crops Journal* 1:69–92.

Pardo, R.D. (1985) 'Forestry for people: can it work?', *Journal of Forestry* 83(12):733–41.

Pardo, R.D. (1993) 'Back to the future: Nepal's new forestry legislation', *Journal of Forestry* June:22–6.

Partap, T. (1991) *Farming Systems and Forestry Related Ecological Aspects.* Himachal Pradesh Environmental Forestry Project. Technical Report to ODA, London.

Partridge, W. (1990) Forest Dwelling People in Bank-assisted Projects', Washington, DC: Environment and Social Affairs Division for Asia, World Bank.

Pathak, A. (1994) *Contested Domains: The State, Peasants and Forests in Contemporary India.* New Delhi: Sage Publications.

Pathan, R.S. (1994) 'Emerging trends in sustainable use plan: tending and harvesting in JFM areas', *Wasteland Newsletter* 9(4).

Pathan, R.S., Arul, N.J. and Poffenberger, M. (1991) *Forest Protection Committees in Gujarat – Joint Management Initiative.* Sustainable Forest Management Working Paper No. 7. New Delhi: Ford Foundation.

Picciotto, R. (1992) *Participatory Development: Myths and Dilemmas.* Policy Research Working Papers WPS 930, Washington, DC: World Bank.

Picciotto, R. and Weaving, R. (1994) 'A new project cycle for the World Bank?', *Finance and Development* 31(4):42–5.

Poffenberger, M. (1990) Joint Forest Management in West Bengal: The Process of Agency Change. Sustainable Forest Management Working Paper No. 9. New Delhi: Ford Foundation.

Poffenberger, M. and McGean, B. (eds) (1994) 'Proceedings of the policy dialogue on natural forest regeneration and community management'. *Research Network Report No. 5*, Honolulu, Hawaii: The Asia Sustainable Forest Management Network, East-West Center.

Poffenberger, M. and Sarin, M. (n.d). Fiber Grass from Forest Land: a Case from North India. Sustainable Forest Management Working Paper No. 10, New Delhi: Ford Foundation.

Poffenberger, M. and Singh, C. (1992) 'Emerging directions in Indian forest policy: legal framework for joint management', *Wasteland News* 7(3): 4–11.

Pokharel, B.K., Gayfer, J.R., Epstein, D.M., Kafle, G. and Tumbahampe, N. (1993) 'Post-formation support for forest user groups', *Banko Janakari* 4(1):80–4.

Poudyal, A.S. and Edwards, E.L. (1994) 'Evaluation of range level planning in Bajhang District'. Report prepared for Community and Private Forestry Division, Kathmandu.

Rahnema, M. (1992) 'Participation'. In W.Sachs. (ed.) *The Development Dictionary.* London: Zed Books Ltd.

Raju, G., Vaghela, R. and Raju, M.S. (1993) *Development of People's Institutions for Management of Forests.* Ahmedabad: VIKSAT.

Rastogi, A. (1995) 'Impact of culture on process of joint forest management in India', *Ambio* XXIV(4):253–5.

Rathore, B.M.S. (1994) 'Options for silvicultural management of regenerating forests under JFM'. Paper presented at the 4th International Congress of Ethnobiology, Lucknow, 17–21 November.

Rathore, B.M.S. and Campbell, Jeffrey Y. (1995) 'Evolving forest management systems, innovating with planning and silviculture', *Wasteland News* XI(1).

Raval, S.R. (1994) 'Wheel of life: perceptions and concerns of the resident peoples for Gir National Park in India', *Society and Natural Resources* 7(4):305–20.

Regmi, M.C. (1978) *Thatched Huts and Stucco Palaces: Peasants and Landlords in 19th Century Nepal.* New Delhi: Vikas Publishing House.

Regmi, M.C. (1982) *Decentralisation Act (English translation)* Nepal Miscellaneous Series Vol. 15/84. Kathmandu: Regmi Research Ltd.

Regmi, M.C. (1984) *The State and Economic Surplus: Production, Trade and Resource-Mobilisation in Early 19th Century Nepal*. Varanasi: Nath Publishing House.

Ribbentrop, B. (1900) (1970) *Forestry in British India*. Republished in 1970. New Delhi: Indus Publishing Co.

Richardson, S.D. (1978) 'Forestry: unique vehicle for development', *Asian Development Bank Quarterly Review* 4: 8–16.

Riddell, R. (1987) *Foreign Aid Reconsidered* . London and Baltimore, MD: James Currey and ODI, Johns Hopkins University Press.

Rizvi, S.S. (1994) 'Managing the forests – herdsman's way', *Wastelands News* IX(4):26–8.

Robinson, P.R. and Neupane, H.R. (1988) 'Preliminary results of the Dolakha private tree survey. In Proceedings, 1st Working Group Meeting on Fodder Trees, Forest Fodder and Leaf-Litter. FRIC Occasional Paper 2/88:36–40,Kathmandu: Forest Department.

Romm, J. (1981) 'The uncultivated half of India. Part 1 and 2', *Indian Forester* 107(1 & 2):1–21 & 69–84.

Rondinelli, D. (1983) *Development Projects as Policy Experiments: An Adaptive Approach to Development Adminstration*. London and New York: Methuen.

Rostow, W.W. (1971) *The Stages of Economic Growth*. Cambridge: Cambridge University Press.

Roy, S.B. (1992) 'Forest protection committees in West Bengal, India: emerging policy issues'. Calcutta: IBRAD (mimeo).

Roy, S.B. (n.d.)'Bilateral matching institution: an illustration in forest conservation'. Calcutta: IBRAD (mimeo).

Roy, S.B., Mukherjee, R. and Chatterjee, M. (n.d.) *Endogenous Development, Gender Role in Participatory Forest Management*. IBRAD Technical Paper No. 3, Calcutta: IBRAD.

Roy, S.B., Mukherkee, R., Roy, D.S, Bhattacharya, P. and Bhadra R.K. (n.d.) *Profile of Forest Protection Committees at Sarugarh Range, North Bengal* IBRAD Working Paper No. 16, Calcutta: IBRAD.

Roychowdhury, A. (1994) 'Serving notice on Indian forests', *Down to Earth* 31 August:5–8.

Roychowdhury, A. (1995) 'The woods are lovely. . .', *Down to Earth* 31 January: 25–30.

Runge, C.F. (1986) 'Common property and collective action in economic development', *World Development* 14 (5):623–35.

Sanwal, M. (1987) 'The Implementation of Decentralisation: a Case Study of District Planning in Agra District, Uttar Pradesh, India', *Public Administration and Development* 7(4):383–97.

Sanwal, M. (1988) 'Community forestry: policy issues, institutional arrangements, and bureaucratic reorientation', *Ambio* XVII(5):342–6.

Sarin, M. (1993) *From Conflict toCcollaboration: Local Institutions in Joint Forest Management*. Joint Forest Management Working Paper No. 14, New Delhi: SPWD.

Sarin, M. (1994) 'Regenerating India's forests: reconciling gender equity with joint forest management'. Paper prepared for the International Workshop on India's Forest Management and Ecological Revival, New Delhi 10–12 February, University of Florida and Tata Energy Research Institute.

Sarin, M. (1995) 'Joint forest management in India: achievements and unaddressed challenges', *Unasylva* 46(180): 30–6.

Sarin, M. and SARTHI (1994) 'The view from the ground: community perspectives on joint forest management in Gujarat, India'. Paper presented at Symposium on Community Based Sustainable Development, IDS, Sussex, 4–8 July.

Sartorius, P. and Henle, H. (1968) *Forestry and Economic Development*. New York: Frederick A. Praeger.

Saxena, N.C. (1990) *Joint Forest Management: a New Development Bandwagon in India?* Rural Development Forestry Network Paper 14d, London: ODI.

Saxena, N.C. (1992) 'Social forestry: why did it fail?', *The Hindu Survey of the Environment*: 35–9.

Saxena, N.C. (1994a) *India's Eucalyptus Craze: The God that Failed*. New Delhi: Sage Publications.

Saxena, N.C. (1994b) 'NTFP – policy issues'. Paper presented at the International Seminar on Management of NTFP, Centre for Minor Forest Products, 12–15 November, Dehra Dun.

Scherr, S., Buck, L., Meinzen-Dick, R. and Jackson, L. A. (1995) *Designing Policy Research on Local Organisations in Natural Resource Management*. EPTD Workshop Summary Paper No. 2, Washington, DC: IFPRI.

Schlich, H. (1922) *Manual of Forestry: 1 Forest Policy in the British Empire*. London: Bradbury Agnew.

Scott, C. and Gupta, J.R. (1990) *Forest Resource Conservation and Development Plan: Shivalik Hills*. Sustainable Forest Management Working Paper Series No. 6, New Delhi: Ford Foundation.

Sen, D. and Das, P.K. (1987) *The Management of People's Participation in Community Forestry: Some Issues*. Social Forestry Network Paper No. 4d, London: ODI.

Shah, A. (1991) 'Participatory afforestation: a challenging opportunity', *Wasteland News* May-July: 23–5.

Shah, S.A. (1975) 'Forestry as an instrument of social change', *The Indian Forester* 101(9):511–15.

Shah, S.A. (1994a) 'Silvicultural management of our forests', *Wasteland News* 9(2).

Shah, S.A. (1994b) 'Reinventing tropical forest management in India', *Indian Forester*, June.

Shankar, K. (1994) Jawahar Rozgar Yojana: an Assessment in UP' *Economic and Political Weekly* xxix (29): 1845–8.

Sharma, A. (1995) 'Lost in the jungle', *Down to Earth* 31 January:34–6.

Sharma, R. (1994) 'Learning from experiences of joint forest management in India', *Forests, Trees and People Newsletter* No.24: 36–41.

Sharma, U.R. (1993) 'Community forestry: some conceptual issues', *Banko Janakari* 4(1):9–11.

Sheikh, A.M. (1989) *The Forest Products Marketing System in Nepal: a Case Study of Urban Areas of Kathmandu Valley.* Kathmandu; USAID.

Shepherd, G. (1992) *Managing Africa's Tropical Dry Forests: A Review of Indigenous Methods.* ODI Occasional Paper, London: Overseas Development Institute.

Shields, D. (1994) *Initial Project Process Development Workshops* Western Ghats Forestry Project Process Support Team Document No. 16, Karnataka Forest Department and the Overseas Development Administration.

Shimizu, T. (1994) 'Community forestry in the Annapurna Conservation Area Project (ACAP), Nepal: comparative study of a sponsored forest management system and an indigenous forest management system'. Unpublished MSc. thesis, Department of Forestry, Wageningen Agricultural University, The Netherlands.

Shiva, V. and Bandyopadhyay, J. (1983) 'Eucalyptus – a disastrous tree for India', *The Ecologist* 13(5):184–7.

Shiva, V. Bandyopadhyay, J. and Jayal, N.D. (1985) 'Afforestation in India: problems and strategies', *Ambio* 14(6): 329–33.

Shiva, V., Sharatchandra, H.C and Bandyopadhyay, J. (1981) *Social, Economic and Ecological Impact of Social Forestry in Kolar,* Bangalore: Indian Institute of Management.

Shiva, V., Sharatchandra, H.C. and Bandyopadhyay, J. (1982) 'Social forestry: no solution within the market', *The Ecologist.* 12(4): 158–68.

Shiviah, M. and Shrivastava, K.B. (1990) Factors Affecting Development of the Panchayati Raj System, Hyderabad: National Institute of Rural Development.

Shrestha, K.B. (1995) 'Community forestry in Nepal and an overview of conflicts', *Banko Janakari* 5(3):101–7.

Shrestha, K.B. and Budhathoki, P. (1993) 'Problems and prospects of community forestry development in the Tarai region of Nepal', *Banko Janakari* 4(1):24–7.

Shrestha, R. and Drona, K.C. (1994) 'Constraints and problems encountered in using some PRA tools'. In Proceedings of the First Nepal Participatory Action Network Workshop, Dhulikhel, 21 January.

Shrestha, B. and Pandey, D. (1989) *Agri-silviculture in Nepal: Study of an Approach to the Settlement of Landless Labourers.* Forestry Research Paper Series No. 14. Kathmandu: HMGN/USAID/GTZ/Ford/Winrock Project.

Siddiqi, N. (1989) 'Towards effective participation: a guide for working with women in forestry'. Technical Note, Nepal-Australia Forestry Project, Kathmandu.

Singh, B. (1953) *Final Report of the 4th Revised Settlement of the Kulu Sub-division of the Kangra District 1945–1952*. Chandigarh: Government of Punjab.

Singh, B., Arora, S.S. and Gupta, J.R. (1993) *Management of Bamboo Forest under Joint Participatory Forest Management in Haryana Shivaliks*. Joint Forest Management Series No. 6, New Delhi: Haryana Forest Department and Tata Energy Research Institute.

Singh, J.S., Pandey, U. and Tiwari, A.K. (1984) 'Man and forests: a central Himalayan case study', *Ambio* 13(2):80–7.

Singh, K. (1992) *People's Participation in Managing Common Pool Natural Resources: Lessons of Success in India*. Institute of Rural Management Working Paper No. 29, Anand: IRMA.

Singh, N.M. and Singh, K.K. (1993) 'Bright outlook in Orissa', *Wasteland News* VII(2).

Singh, S. (1990) 'People's participation in forest management and the role of NGOs and voluntary agencies'. Paper prepared for the Seminar on Forestry for Sustainable Development, Asian Development Bank, Manila, Philippines.

Singh, S. (1991) 'An Indian Government initiative: people's participation in forest management', *Sustainable Development* 1(1).

Singh, S. and Khare, A. (1993) 'People's participation in forest management', *Commonwealth Forestry Review* 72(4):279–283.

Soussan, J., Allsop, N. and Amataya, S. (1992) 'An evaluation of of Koshi Hills Community Forestry Project'. Atkins, Land and Water Management Consultants, for SEADD, ODA, London.

Soussan, J., Gevers, ELS, Ghimire, K., and O'Keefe, P. (1991) 'Planning for sustainability: access to fuelwood in Dhanusha District, Nepal', *World Development* 19 (10):1299–314.

SPWD (1984) 'Hill resource development and community management. Lessons learnt on micro-watershed management from cases of Sukhomajri and Dasholi Gram Swarajya Mandal. Report submitted to the Planning Commission's Working Group on Hill Area Development, SPWD, New Delhi.

SPWD (1992) 'Joint forest management: concept and opportunities'. In Proceedings of the National Workshop at Surajkund, August. New Delhi: SPWD.

SPWD (1993) *Joint Forest Management Update*. New Delhi: Society for Promotion of Wastelands Development.

Srivastava, B.P. and Pant, M.M. (1979) 'Social forestry in India', *Yojana*, July

Stainton, J.D.A. (1972) *Forests of Nepal*. London: John Murray.

Stebbing, E.P. (1927, 1929) *The Forests of India*. London: John Lane, volumes 1–4.

Stewart, M. (1987) *A Guide to the Mapping of Forest and Land Use Types in the Koshi Hills, Eastern Nepal*. LRDC P-180, Chatham, UK: NRI.

Stiller, L. (1973) *The Rise of the House of Gorkha: A Study in the Unification of Nepal 1768–1816*. Kathmandu: Ratna Pustak Bhandar.

Streeten, P. and Bucki, S.J. (1978) 'Basic needs some issues', *World Development* 6(3):411–21.

Subedi, B.P., Das, C.L.and Messerschmidt, D.A. (1991) *Tree and Land Tenure in the Eastern Nepal Terai.* IOF/Yale/IRG/USAID, Pokhara, Nepal: Institute of Forestry.

Swallow, B.M. and Bromley, D.B. (1994) 'Institutions, governance and incentives in common property regimes for African rangelands', *Environmental and Resource Economics.*

Talbott, K. and Khadka, S. (1994) *Handing it Over: an Analysis of the Legal and Policy Framework of Community Forestry in Nepal.* Washington DC: World Resources Institute.

Tamang, D. (1990) *Indigenous Forest Management Systems in Nepal: A Review* Research Report Series No. 12. Kathmandu: Ministry of Agriculture/Winrock International.

Tamrakar, P.R. (1993) 'Management systems for natural *Schima/ Castanopsis* forests in the Middle Hills of Nepal', *Banko Janakari* 4(1):57–62.

Tamrakar, S.M. and Nelson, D.V. (1991) 'Potential community forestry land in Nepal. Part 2'. Kathmandu: Community Forestry Development Project.

Tewari, K.M. (1984) *Social Forestry in India* Dehra Dun: Natraj Publishers.

Thakur, P. (1984) 'Forestry in India: its conservation and planning', *Indian and Foreign Review,* June.

Thomas-Slayter, B.P. (1994) 'Structural change, power politics, and community organisations in Africa: challenging the patterns, puzzles and paradoxes', *World Development* 22(10):1479–90.

Thompson, M., Warburton, M., and Hatley, T. (1986) *Uncertainty on a Himalayan Scale.* London: Ethnographica.

Tiffen, M., Mortimore, M. and Gichuki, F. (1994) More People Less Erosion: Environmental Recovery in Kenya. Chichester, UK: John Wiley.

Timsina, D. and Poudel, B. (1992) 'Farmer participatory approach in identifying gender issues in agriculture and forestry related activities in Jhapa, Nepal'. Paper presented at 12th Annual Farming Systems Symposium, Michigan State University, Lansing, MI.

Tinker, I. (1994) 'Women and community forestry in Nepal: expectations and realities', *Society and Natural Resources* 7(4): 367–81.

Troup, R.S. (1940) *Colonial Forest Administration.* Oxford: Oxford University Press.

Tucker, R.P. (1982) 'The forests of the Western Himalayas: the legacy of British colonial adminstration', *Journal of Forest History,* July: 112–23.

Tucker, R.P. (1983) 'Economic impacts on the Himalayan forests of Punjab and the United Provinces under British colonial rule, 1815–1914'. In R.P. Tucker and J.F. Richards (eds) *Global Deforestation and the Nineteenth Century World Economy.* Durham, NC: Duke University Press.

Tucker, R.P. (1984) 'The historical context of social forestry in the Kumaon Himalayas', *The Journal of Developing Areas* 18(April):341–56.

Tucker, R.P. (1986a) 'The British colonial system and the forests of the Western Himalayas, 1815–1914'. In R.P. Tucker and J.F. Richards (eds) *Global Deforestation in the Nineteenth Century World-Economy*. Durham, NC: Duke University Press.

Tucker, R.P. (1986b) 'The evolution of transhumant grazing in the Punjab Himalaya', *Mountain Research and Development* 6(1):17–28.

Tumbahampe, N. (1994) 'Dhaulagiri joint investigation report'. Report NUKCFP, Kathmandu.

Umans, L. (1993) 'A discourse on forestry science'. *Agriculture and Human Values* 10(4):26–40.

Unnikrishnan, P.N. (1994) 'Joint forest management plan for Kerala'. Mimeo report prepared for course on Forest, People and Participation, University of Wolverhampton.

Uphoff, N. (1986) *Local Institutional Development: An Analytical Sourcebook with Cases*. West Hartford, CT: Kumarian Press.

Uphoff, N. (1992) *Local Institutions and Participation for Sustainable Development*. Gatekeeper Series No. 31. London: IIED.

Uphoff, N. (1993) 'Grassroots organisations and NGOs in rural development: opportunities with diminishing states and expanding markets', *World Development* 21(4):607–22.

Varalakshmi, V., Vijh, R. and Arora, S.S. (1993) *Constraints in the Implementation of Joint Participatory Forest Management Programme – Some Lessons from Haryana*. Joint Forest Management Series No. 12, New Delhi: Haryana Forest Department and Tata Energy Research Institute.

Venkateswaran, V. (1994) 'Managing waste: ecological, economic and social dimensions', *Economic and Political Weekly* 29(45 & 46):2907–11.

Vijh, R. and Arora, S.S. (1993) *Economics of Rope Making under Participatory Forest Management*. Joint Forest Management Series No. 9, New Delhi: Haryana Forest Department and Tata Energy Research Institute.

Vira, S. (1993) 'Joint forest management and nomadic groups – the potential for conflict'. Paper prepared for the Society for Promotion of Wastelands Development, New Delhi.

Voelker, J.A. (1897) *Report on the Improvement of Indian Agriculture*. Calcutta: Government Press.

Von Maydell, H. (1977) 'The contribution of forestry to regional development in the Sahel', *Applied Sciences and Development* 10:164–74.

Wade, R. (1988) *Village Republics: Economic Conditions for Collective Action in South India*. Cambridge: Cambridge University Press.

Warren, S. (ed.) (1992) *Gender and Environment: Lessons from Social Forestry and Natural Resource Management*. Ottawa: Aga Khan Foundation.

WCED (1987) *Our Common Future*, Report of the World Commission on Environment and Development. London: Zed Books.

Weber, T. (1988) *Hugging the Trees: the Story of the Chipko Movement*. New Delhi: Viking.

Webster, N. (1990) *Panchayati Raj and the Decentralisation of Development Planning in West Bengal: A Case Study* CDR Project Paper 90.7, Copenhagen: Centre for Development Research.

Wee, Ai-Chin and Jackson, B. (1993) 'The forest beat – a starting point for integrating participatory planning, budgeting and monitoring in the forestry sector of Nepal'. Paper presented at regional seminar on Forestry Management for Sustainable Development, 4–9 October, Kandy, Sri Lanka.

Westoby, J. (1962) 'Forest industries in the attack on underdevelopment', *Unasylva* 16(4):168–201.

Westoby, J. (1975) 'Making trees serve people'. *Commonwealth Forestry Review* 54(3/4):206–15.

Westoby, J. (1978) 'Forest industries for socio-economic development'. Paper presented at the Eighth World Forestry Congress, Jakarta, Indonesia.

Westoby, J. (1987) *The Purpose of Forest: The Follies of Development*. Oxford: Basil Blackwell.

Westoby, J. (1989) *Introduction to World Forestry*. Oxford: Basil Blackwell.

Wiersum, K.F. (1986) 'Social forestry and agroforestry in India'. Report No. 456, Department of Forest Management, Wageningen Agricultural University, the Netherlands.

Wollenberg, E. and Hobley, M. (1994) *Experiences and Advances in the Development of Community Forestry Programming: a Review of Ford Foundation Supported Community Forestry Programs in Asia*. Report prepared for Ford Foundation, New York.

World Bank (1978) *Forestry Sector Policy Paper*. Washington DC: World Bank.

World Bank (1985) *India: National Social Forestry Project*. Staff Appraisal Report. Washington DC: World Bank.

World Bank (1991a) *The Forest Sector: A World Bank Policy Paper*. Washington DC: World Bank.

World Bank (1991b) 'Technical assistance to the Terai Forestry Project, Nepal'. Terminal statement, report to the FAO and to the World Bank.

World Bank (1994) *Participation Sourcebook: Technical Paper, Forestry*. Washington DC: World Bank.

World Bank (n.d.) *A Strategy for Asian Forestry Development* Washington, DC: Land Resources Unit, Asia Technical Department, the World Bank.

Wright, H.L. (1917) 'Report on the Mandi Forest Settlement'. Mandi State, India.

Young, D. (1994) *Community Forestry Impact Model: Report of Fieldwork in Bhojpur and Dhankuta*. Project Report No. 4, Kathmandu: NUKCFP.

Glossary

Ban goswara Forest office

Ban janch Forest inspection office

Ban karyojana Forest working plan

Bartan Forest rights in Mandi and Kullu districts, Himachal Pradesh

Bartandars Forest right-holders

Bhari Load generally carried in a large basket

Bidhan Written constitution

Birta A grant of land to a noble as a reward for a service rendered to the state. This led to the emergence of *birta* land tenure. It was usually both tax free and heritable, and had no set time limit. It was valid until it was recalled or confiscated (Regmi, 1978)

Chitadar Watcher

Chowkidar Village watcher

Dahay Basket

Dittha Local level functionary

Doko Bamboo basket

Falia Hamlet

Gaja-patra Locally agreed written list of rules

Gram sabha Village committee

Gram vikas mandal Village development committee

Guthi Endowment of land or other property for a religious or philanthropic purpose

Jagir A grant of land to a government employee (civil or military) in lieu of salary. This led to the emergence of *jagir* land tenure. The *jagir* land grant was also tax free but remained valid only as long as the concerned person served the government (Regmi, 1978). Such a practice was also prevalent in Mandi State.

Jagirdar Person receiving land as *jagir* (Regmi, 1978)

Jamabandi Lists of rightholders in Mandi and Kullu

Jimmawal Local functionary during the Gorkhali and Rana periods (see *talukdar* for further information)

Khoriya Seasonal plots in forest usually for millet cultivation

Kipat Ancient type of communal land tenure, applied to both cultivated and forested land. Under this system a community was granted land by the King in recognition of the land's traditional communal tenure. On *kipat* lands, the community (community leader) gave individuals the right to till certain areas and to collect forest products from other areas (Regmi, 1978)

Kothi Revenue unit in Kullu district

Lauro Staff

Mana Nepali traditional unit of volume, 1 *mana*=0.31 to 0.44 kg depending on the type of grain.

Mana-pathi In the context of common property forest management, a system adopted in which users of the forest decide to appoint forest watchers and each household from within the user group contributes a certain amount (usually in volumetric measure of *mana* or *pathi*) of grain to pay the watcher as salary

Mukhiya Local leader

Nallah Stream

Nautor It is an ancient right, now not permitted, whereby landless people are allowed to break fresh agricultural land in common land areas. The land is allocated to the landless by village elders, usually on undemarcated (Class III) land in Kullu, Himachal Pradesh.

Negis Locally appointed officials in Mandi state

Neja Prayer flag

Nistar Rights to use of forest products (Madhya Pradesh)

Panchayat Lowest administrative and political unit encompassing a number of contiguous villages and settlements, consists of elected members headed by a chairman.

Pathi Nepali traditional measurement unit of volume, 1 *pathi* = 2.5 to 3.5 kgs depending upon the type of grain

Praman patra Certificate of ownership

Rakam Compulsory labour obligations which a farmer rendered to government and later also to the *birta* owners on a regular and inheritable basis (Regmi, 1978)

Rakhas Local forest guards (pre-1860s) in Mandi and Kullu

Rato pheta Red turban

Ropani Nepali unit of land area, mostly used in the hills; 1 ropani = 0.05 ha

Sanad Rule or decree

Satyagraha Civil dissent, uprising against the colonial authorities

Shamlat Common lands

Shikari Hunter

Siyan Demarcated forest (Mandi State)

Talukdar A local functionary (usually a hereditary position) of the state who existed from the time of the Gorkhali rulers (1769–1846) until just after the end of the Rana regime in the 1950s. According to Regmi (1978) the term *talukdar* is a generic term covering local headmen such as the *mijar* in some Buddhist groups and the *mukhiya* or *jimmawal* in other contexts. The main function of the *talukdar* was to collect land tax for the state from the local community members living in the taluk for which he was responsible. According to Mahat (1985) the *talukdar* also had the responsibility for controlling access to the forests and for distributing forest products.

Talukdari system In the context of the common property forest management, the system refers to the management of forests under the control or direction of the *talukdars* (see *talukdar*)

Tans Huts erected in fields from which to watch the crops and protect from animals

Taungya A system of tree management on state land (usually) where farmers are permitted to plant agricultural crops for the first few years whilst the tree plantation is establishing itself. When the tree cover has closed farmers move from that area to another where the plantations are to be established. The idea is that farmers gain some short-term benefits from the land in the form of agricultural crops whilst in return the young trees are protected. The system was first established in Burma, in the nineteenth century, in teak plantation areas.

Tengapalli Rotational patrolling system (Orissa)

Thaches Upland grazing areas

Van panchayat Forest panchayat

Vana samrakshana samiti Forest Protection Committee

Vara Rotational patrolling system (Gujarat)

Zamindars Landlords during the colonial period

A Appendix: From Passive to Active Participation Group Exercise

Purpose:

To experience a passive group meeting from the perspective of different participants.

Each group has been given different roles, and does not know what the other groups have been told to do. The exercise can be adapted to accommodate different numbers of participants. An example is presented based on a group of more than 30 participants.

Background:

Village name:	Dungagaon
Population:	200
Forest area:	Protected state forest (rights to collect dry firewood and some non-timber forest products)
Village Forest Committee:	Set up six months before at the initiative of the project
Project:	Forest Development Project through joint forest management
Donor:	Bilateral donor (grant)

Group 1 (16 people)

Decide among yourselves who is going to play the part of the poor men, and the women.

Roles:

Poor men

Important points.

5 out of 10 poor men are the tenants of the village elders (one man was elected on to the village forest committee).

The other 5 owe large amounts of money to the village elders.

None of them have sufficient land of their own to supply all their tree product needs.

Women

2 out of 6 of the women are young daughters-in-law of village elders.

1 woman is a widow with a large amount of land, and a son who works in the city and sends money home to her regularly (she is a member of the village forest committee).

2 women are wives of two other participants in the meeting. They have sufficient agricultural land to provide all their firewood and fodder needs. (One of these women is a member of the village forest committee).

2 women are headloaders. They have no land of their own and rely entirely on free access to the forest to collect dead wood for sale in the local market. They have no other source of income or livelihood.

Background

It is the afternoon and you are very busy. You have just been rounded up from the fields and from your homes where you were working by several of the village elders. They have told you that important visitors have arrived from the Forest Department and from England, and you have to come to a meeting.

You say you are too busy, but they say you must come or else there will be trouble. You are not told what the meeting is about, or how long it is going to continue. The women headloaders say that they have to go to sell their firewood, otherwise they will not have any money to buy food for the evening meal. The elders say that they must come to the meeting; if they do not they may get into trouble with the Forest Department, and questions will be asked about the firewood they are selling: is it really dead, or is it green?

In your small groups decide:

1. who is going to play which person
2. how you are going to behave in the meeting
3. make sure you consider all the background information given about the circumstances of the meeting and who you are.

Seating arrangements:
Arrange yourselves as shown in the diagram.

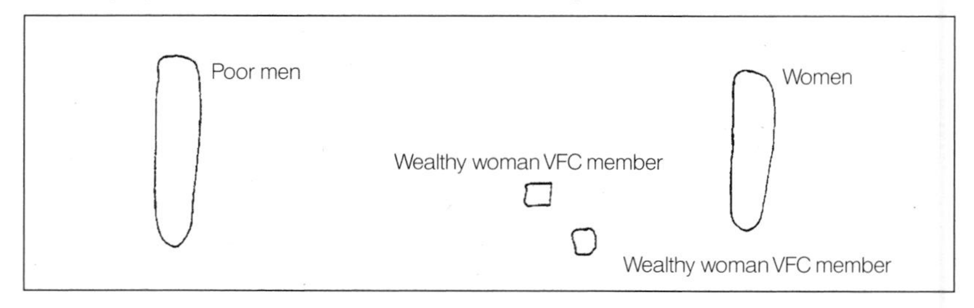

Group 2 (10 people)

Village elders:
One ex-military man (has a good pension and knows something about the world and how to deal with foreigners). He is chairman of the village forest committee.

2 wealthy fathers-in-law whose daughters-in-law are also at the meeting. They have joint households with autocratic decision-making systems. They are both members of the village forest committee.

2 very elderly men with good memories of how the forest used to be.

Village leaders
(all men)
One young political activist who wants to get into local government.

A local trader (thought, by the other villagers, to be involved in illegal timber smuggling). He is a member of the village forest committee.

3 wealthy farmers (money-lenders and landowners).

Background

You were told an hour before the DFO and the foreigner arrived that there was to be a meeting to discuss joint forest management activities in the village.

You have told the women and poor men to come to the meeting. You have not told them anything about why the meeting has been called, mainly because you do not know, but also because you do not think they need to know why.

The trader is keen to be at the meeting because he wants to make sure that no decisions are taken that may affect his trade.

The political activist is trying to get on to the village forest committee and become its chairman, the current chairman is from a different party, and is trying to buy votes for his party by giving his party's supporters better access to the forest.

In your small groups decide:

1. who is going to play which person
2. how you are going to behave in the meeting
3. make sure you consider all the background information given about the circumstances of the meeting and who you are.

Seating arrangements:

Arrange yourselves as shown in the diagram.

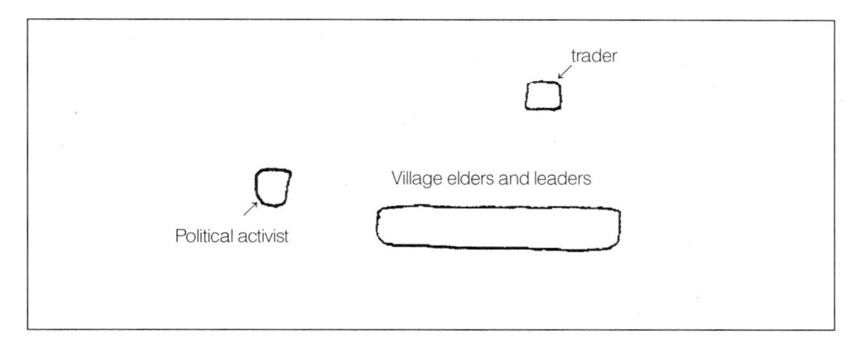

Group 3 (5 people)

Forest guard
Forester (village is in his section)
RFO
DFO
Expatriate forest consultant (optional)

Background:

The DFO was told by the CF the day before that a consultant from the project was coming to his division to meet some of the villagers involved in developing joint forest management. The DFO has never met the consultant before and is not clear what the consultant will want to do. He has been told that the consultant will arrive at his office at 2 p.m. after lunch with the CF at the forest rest house.

The DFO phones the RFO of a range where there are some JFM villages and summons him to the office on the morning of the visit.

The RFO is summoned by the DFO for a meeting, but he is not

told why because the DFO is not clear and also the DFO does not think it is necessary to tell the RFO. They have a meeting in the morning and the DFO tells the RFO to let Dunga village forest committee know that they will be coming with a foreigner in the afternoon to meet the villagers and talk about JFM.

The RFO hurries back to his range office, arrives back at lunch-time and summons the forester in whose area is the JFM village. The latter arrives and is immediately despatched to the village to say that in two hours time the foreigner is coming to 'inspect their work'.

The RFO instructs the forester to tell the forest guards they must also be present and in clean uniform because an important foreigner is coming to inspect the JFM operations.

Meanwhile the foreigner is happily eating a large lunch at the rest house, blissfully unaware of the trouble he is causing. He has been in the country for two days and has only another four days to finish a review of participatory forest management. This is a tight schedule but he thinks it is better not to give people time to organise meetings with villagers so that they don't go to any special trouble.

In the jeep five minutes before they reach the village, the foreigner decides that he and the DFO should have a list of questions for the meeting, so that they can collect as much information as possible.

In your small groups decide:

1. who is going to play which person
2. how you are going to behave in the meeting
3. make sure you consider all the background information given about the circumstances of the meeting and who you are.

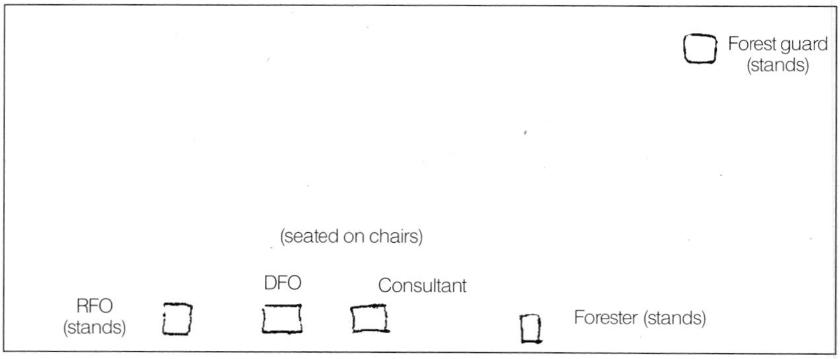

Forest guard (stands)

(seated on chairs)

RFO (stands)

DFO

Consultant

Forester (stands)

Seating arrangements:

Arrange yourselves as shown in the diagram.

Group 4

Observers of the social dynamics of the meeting

Your task is to note what is happening in the meeting who is speaking and who is not.

What are your recommendations for changing the meeting to make sure that all the forest users at the meeting have an opportunity to put forward their points of view?

At the end of the meeting you will ask each group to report back what they felt about the meeting, what were the good points, and what were the bad points. How would they like to have changed the way the meeting was conducted?

Final Exercise

An interest group analysis sheet can then be filled in which will allow analysis of how the different groups view the project and its interventions in their lives. (The format provided is an example that

Interest Group	Interest in the forest	Consequence of the project on each interest group		Project Strategy
		+	-	

can be adapted for use with this exercise.) The last column is to be filled out in consultation with all the groups and indicates the strategy the project should adopt to alleviate any problems or build on constructive experiences.

B Appendix: Group Formation Process

Purpose:

The case study presented below can be used to discuss ways in which to set up user groups.

How not to set up a user group: case study

First of all the DFO looks at the budget, and realises that there are still funds outstanding for the establishment of user groups. He rings the bell on his desk and soon a ranger turns up, shyly and stiffly he greets the DFO. The DFO orders him to form the number of user groups stipulated in the budget.

The ranger goes off on his bicycle and heads towards a village. He knocks at the door of the chairman of the Village Development Committee and orders:

Ranger:	'I have come from the forest office to form a user group committee. We need 12 to 15 people from your village on this committee.'
Chairman:	'Do they get paid?'
Ranger:	'No, but they get a letter pad, stamp and sign board.'
Chairman:	'This is very difficult, let's call a meeting.'

During the next few hours, the village men gather under the **pipal** tree, drink tea, and form the user group committee.

Village men:	'There are no salaries . . .'
Village men:	'Let's put him on the committee, and him . . .'

They make a list of people, some of them present at the meeting, others not. When some are told to join the committee they refuse and refer to others. Finally they succeed in drawing up a list of user group committee members. The ranger asks them to sign the paper and cycles back to his office.

The work is finished and another 'paper' committee has been formed.

In some cases there is some follow-up to this first meeting.

The new user group chairman puts on his best clothes, combs his hair and walks to the forest office. He travels the last kilometre by rickshaw. He meets the ranger and together they arrange for a letter pad, stamp and sign board. They even plan to have a plantation in the village development committee area because the ranger has some money left in the budget for a plantation.

The chairman returns to his village, proud of his achievements and satisfied with his plans for the community forest. His social status in the village improves as he distributes work opportunities and employs a local forest watcher.

The ranger and the chairman work well together, so well, in fact, that the ranger can manage the plantation without having to go to the village a single time. In the evenings they meet and drink beer and eat meat in the bazaar and discuss future village forest activities.

Finally, a volunteer is attached to the District Forest Office, and the DFO suggests that it would be helpful if the volunteer were to meet some of the user group committees and provide support to the ranger with user group formation.

The volunteer has read all the documentation about the community forestry programme and checks if all the interest groups are involved in the user group. The villagers quickly add a woman to the committee (she is a wife of one of the committee members). At the insistence of the volunteer, some low-caste villagers are appointed to the committee. They do not understand their role on the committee and are obliged to agree with everything their landlord says (he is also a committee member).

The volunteer looks for new ways of reaching the 'silent majority'. Attending an adult literacy class she realises that the major proportion of the students are from the poorer groups in the village. This provides an important forum in which to develop their understanding of the community forestry programme and the benefits to be obtained from it.

Source: Dumortier, 'Why foresters get involved in literacy programmes' *The Woodpecker* (the magazine for VSO Nepal forestry volunteers).

Discussion questions

Analyse this case study and describe the problems with the methods chosen to set up this group. What are the implications of this for the future sustainability of this group?

Are there ever circumstances in which it is appropriate to set up groups in this way? If there are, describe what these are.

C Appendix: Who are the Stakeholders?

Purpose:

The purpose of this exercise is to provide an example, as given below, of the types of stakeholders to be considered in the design and implementation of a forestry project. This could be used as a basis from which to design a list of stakeholders for each participant's country. As a follow-on to this further group discussion could be focused on the potential impact on each of these stakeholder groups of the introduction of a participatory forestry project.

Types of Stakeholders	Composition & sensitivities to changes in forestry project
Central Forestry Department	Major policy- and decision-makers; key actors in forestry reforms
Other Government Agencies	Vital in policy- and decision-making and inter-agency co-ordination
Sub-national Forestry Agencies	Administrative unit; in charge of project implementation and budget releases; monitoring of project outputs; dependent upon central department for priorities
Forestry Project Agencies	In charge of day-to-day operations of the project; sometimes operate as autonomous units
State Forestry Corporations	In some countries, state forestry corporations are the major project implementors or they play important roles in marketing and processing forest products
Local Governments	Nominal involvement in project operations; sometimes informed of the project
Local Communities	May be defined as an administrative unit (political); as a forestry unit (farm, woodlot, plantation); or as a social unit (indigenous, migrant)
Affected Populations	Defined by village or household; may refer to populations (resident or non-resident) affected by the forestry project's outcomes
Disadvantaged Populations	Groups of people negatively affected (or marginalised) by the outcomes of the forestry project (e.g. displaced indigenous peoples, women, landless labour)
Small Groups	Examples include exchange or reciprocal labour groups
Associations and Co-operatives	Larger social groupings with well-defined rules and regulations governing participation in group activities; benefit and cost sharing of forestry outputs
Forest User Industries	Large-scale (e.g. pulp mills; rayon factories; lumber companies) or small-scale forest industries (e.g. furniture makers) that influence processing and marketing
Women's Groups	Organised (e.g. group-based fuelwood collection) or unorganised; important in productive or household-related activities
Non-Resident Interest Groups	May be small or large groups of people that are organised (e.g. grazing cooperatives; medicinal herb collectors) or unorganised (e.g. consumers of fuelwood)
Community-based NGOs	Non-governmental groups in (or near) the project area; may be religious based, political, educational; specialised in community development work
Forestry/Environmental NGOs	Non-governmental groups generally based outside project area; may play advocacy roles
Other Donors	Bilateral or multilateral donors providing external and additional financial assistance to forestry projects

Source: adapted from World Bank, 1994

D Appendix: Challenging Assumptions

Gyanendra Kafle and Netra Tumbahampe (1993)

Purpose:

Most people make assessments of situations based on a series of assumptions which often remain unchallenged. The purpose of this exercise is to provide an example of how to get participants to challenge some of the basic tenets of their understanding.

Method

The statements are in pairs, and are split into the different components of community forestry. One statement emphasises a positive aspect of community forestry and the other states the opposite position. Each statement should be written on a separate piece of card or paper and each component in a different colour, to highlight the phases of activity.

Read out each pair of statements and ask participants to distinguish which is positive and which negative. Encourage discussion. When everyone agrees put the positive card on one side of the board, and the other on the other side.

1. It is impossible for a large number of people to reach consensus, as there will be too many different views	1. Though there may be many different views, there is also some common ground on which everyone can agree
2. Women, scheduled caste people and the poor should be encouraged to put forward their ideas in meetings and assemblies. Their voices should be heard because they will make practical suggestions about matters that affect their daily lives	2. Those with high social status should lead and tell everyone else what to do. Everyone should follow what they say. It is not right to argue and discuss with them because they know what is right
3. If consensus is difficult to reach in a meeting or assembly, the majority decision should be upheld	3. Discussion should be continued in meetings and assemblies until consensus is reached. All problems have solutions
4. Women are engaged in household tasks. They can neither say anything in meetings nor do they have any skill to do so. In meetings related to forestry, women are not required to attend	4. Women are generally honest and not involved in trying to manipulate discussions to their advantage. Since women are most involved in forestry-related activities, it is essential to involve them in discussions and meetings related to forestry
5. All the rules contained in the operational plan should be prepared through discussion in the users' assembly. In doing so the views of everyone should be taken into account and consensus should be reached. The committee should be elected by the assembly, with the mandate to implement the operational plan on behalf of the users' assembly	5. It is not practical to discuss and make decisions on all aspects of the operational plan in the assembly. It is the committee members who should decide what to include in the plan. They are the ones who know who should be asked to do what, who should be punished, who should be given privileges, etc.
6. Discussion creates conflicts which ultimately lead to social disintegration	6. Discussion leads to concrete and practical decisions
7. Poor Sante! He has only a small piece of unproductive land with a few useless trees. How can he manage for one year with only five bundles of fuelwood from the community forest, as his share?	7. Don't worry about Sante or anyone else. If everybody takes care of the forest equally, everybody should also get equal shares of the products
8. If there is a debate about who should be on the committee, they should be elected by popular vote. The minority should agree with the majority. This is a multi-party system, we should also act in the same way as this system	8. It is not practical to politicise matters related to activities such as use of roads, trails, forests and household activities. There should be mutual understanding and co-operation. If a vote is taken there is a possibility of those who are in the minority withdrawing their support from the majority agreement
9. Writing the operational plan, amending it and taking it to the DFO for approval should be done by the rangers. This is not work to be done by illiterate villagers. Rangers know better than villagers, as they are the technicians	9. The operational plan is prepared through discussion with all the users. It is amended when necessary. The committee can take the plan to the DFO for its approval. People know their situation and needs better than technicians. They should make the rules on their own, with advice, if requested, from the technicians
10. A forest located in one VDC should not be used by people of other VDCs because this is the property of the VDC. Whoever does not have forests in their VDC/wards will have to buy forest products. This will help to increase the group's funds	10. All those for whom the forest is accessible can be members of the user group. Political and administrative boundaries cannot be criteria for user identification because people have traditionally been using forests which are accessible to them

11. Seasonal users should not be given access to the forest; they should use forests closer to their own areas	11. Seasonal users should be given negotiated access to the community forest. If the local community is investing in protection of the forest and thus the seasonal users' products, some form of payment should be agreed to be paid by the seasonal users to the local protectors for the costs they incur in protection
12. I am the chairperson of 'ABC' user group committee. Our group have planted bamboo, pine and amliso seedlings. Each group member collected a bamboo rhizome from their own farm. Everybody participated in the plantation work with great enthusiasm	12. I am the chairperson of 'XYZ' user group committee. Though the villagers had not agreed, I made them plant cardamom. I am also thinking of making a park inside the forest with the help of the villagers in the winter. I have also got the thorny area cleared. The people can not understand; some did the work, some did not even turn up. It is difficult to make people understand about the importance of forests
13. The forest cannot support landless people. It can scarcely support the needs of the next generation. There is an urgent need to find alternative livelihoods for landless people. Since everybody needs forests, it is wiser to use only the forest income rather than to exhaust the capital itself	13. Landless people should be resettled in the forest area. In doing so, they will have land to grow food, timber for building their houses and fuelwood for cooking. Forests will not be destroyed by allocating small areas for settlements
14. The forest should be managed so that it produces fuelwood, timber and fodder on a continuous and sustainable basis	14. All the large trees should be felled immediately and distributed amongst all users
15. Are villagers supposed to grow and plant seedlings in addition to taking responsibility for looking after the forest? What are the rangers going to do then? Since the rangers have studied forestry, they know all about it. How can villagers, who do not know anything about forestry, do the protection and development work? The government employees should take care of the forest	15. When the forest is handed over to villagers, the use rights are also given to them. The care of the forest will be their responsibility
16. The committee, operating within the limits of its authority, can take action against those who violate the rules prepared by the users' assembly. The committee is empowered to take action if the rules have been agreed by consensus	16. The committee will have difficulty taking action against those who violate rules; the committee should therefore refer the cases to the DFO for necessary action
17. Community forests should be completely protected. There should be neither removal of forest products nor grazing. Forests should be allowed to become a habitat for wild animals, and communities should be rewarded for such forests	17. The forest should be managed to remove dead, diseased and dying trees. Thinning, singling and pruning should be encouraged and products from these operations should be distributed amongst the users.
18. The project should employ watchers to take care of the forest, Villagers cannot spare the time to watch the forest voluntarily	18. Whoever (person or institution) owns the property should take care of it. If a project pays watchers to protect the forest, the forest is considered to belong to that project. Similarly if the government employs watchers, the forest is known as government forest. Since the users own the forest, they are the ones who should look after it

19. After the forest has been handed over to user groups there is no work for HMGN staff. No cases are filed. They do not need to do anything, as nursery and plantation work is also carried out by the groups. So forestry staff are paid their salary for no work

19. The handover of forests is only the beginning of the work for forestry staff. This is, of course, the end of their traditional role, but it is a new phase where their role is one of social and technical advice, and monitoring

Note

These points are presented as examples of possible assumption statements that could be used in a group discussion. In different social and political contexts different assumptions may prevail and these should be used as the basis for an 'assumption-busting' session.

E Appendix: Participatory Forest Management Planning Exercise

Jeff Campbell

Purpose of exercise:

To stimulate community discussion about different management options. In this case the exercise can be used with a group of forest officers or other practitioners to indicate how it is possible to work with villagers to identify their different management requirements. The group could be split into villagers and Forest Department staff. The villagers should be given a briefing document indicating how the forest has been used over the last 20 years. The group representing the Forest Department should not be given this briefing document.

Experience in the field indicates that much can be learned from such an exercise and much more responsive forest management strategies can be evolved which maximise multiple objectives.

Step 1 Sketch out a rough historical transect which reflects major changes over time in forest vegetation and product availability. Extract the important time periods in the history of the forest and its use: e.g. a period when the forest was in excellent condition (perhaps 20 years ago); time of a major event such as a large harvest or other disturbance; the period of maximum use and degradation; period of initial response following protection. Table E.1 provides an illustration of this exercise.

Step 2 Make a list of the most important forest products identified by the users. These are taken as forest management objectives. Additional objectives of the Forest Department, if different, may also be listed. Name these, or make a drawing of each product, across the top of a large sheet of paper or on the ground. On the vertical axis list the major time periods which emerged from step 1.

Step 3 Ask group members to compare the yields and availability or incidence of collection or use of each product over time. This can be done by assigning 10 stones to each time period for each product and evolving a ratio. For instance, if fodder is being discussed, ask participants to indicate the quantity of fodder (out of 10) that was available 20 years ago, when the forests were in excellent condition, 5 years ago when the forests were totally degraded, and now after 5 years of protection.

Step 4 Discuss the ratios that evolve. Some products may have been abundant or easily accessible, like fodder grass for open grazing, or leaves for plates or **bidis** (local cigarettes), when the forest was more open and degraded, while others like large timber or fruit or mushrooms might be more abundant, depending on the age of the forest, or the soil conditions. Make notes on the ecological issues which emerge pertinent to why certain products are more abundant at certain times or in special conditions.

Step 5 Create another line for demand for each product (do not put a limit on the number of stones, but specify that each stone has the same value as earlier). This should generate a discussion about the gap between supply and demand over time. Discuss how long the current forest needs to be protected in order to meet demands, or to return to the condition that it was in over twenty years ago. If it is quite clear that demand far exceeds supply then discuss what ratio could be sustainably supplied from the forest, and where alternative supplies could come from, and how this could be calculated and monitored.

Step 6 Create another line with boxes for management treatments under each product and list the kind of management needed for each product: e.g. some trees should be kept in a bushy condition, or in small diameter classes for the collection of leaves, for fuelwood small diameter stems of coppicing species in the 2–5-year-old age class may be considered ideal, for agricultural implements intermediate-sized trees of certain species are needed, for seed and fruit-bearing trees a minimum age and size is required, for household roofing and other timber needs, large diameter trees are necessary. For mushrooms, moist soil conditions with plenty of organic matter and humus may be needed; for biodiversity preservation some part of the forest should remain untouched. Ask people to sketch what the ideal forest stand would look like and illustrate the production of each objective/product in a separate box. This should generate an interesting discussion on the trade-offs between different prescriptions and different needs – and perhaps generate another priority ranking exercise for different management options. A number of possible strategies may emerge for sequencing different silvicultural operations, for zoning and creating a felling series, for product or species-based management of individual trees and shrubs, for enrichment planting, and for protection.

Step 7 On a rough sketch map of the forest chalk out the zones most suitable for different forest management strategies and discuss the implications of different approaches. An idealised transect could also be drawn which illustrates prioritised proportions of the forest under different treatments.

Step 8 Conduct the same exercise separately with different user groups, men and women, and different occupational groups. Share the results in a general meeting and get comments from Forest Department officers. Options which people suggest under different objective headings could be discussed and options which foresters consider silviculturally viable could be reviewed publicly. This requires discussion of the ecological needs of different species, and the overall structure and sequencing of the forest operations under discussion, which are complex issues. Discuss the practical issues of the implementation of different intermediate and final harvest operations, who would do the operations, what are the economies of scale (when undertaking harvests of small isolated coupes)? How does the forest management plan link with neighbouring forest areas or larger contiguous forest?

Table E.1 Participatory forest management planning matrix

Product/ Objective							Biodiversity	Sustainable
20 years ago when the forest was dense	o o o o o o o o o o	o o o o o	o o o o o o o o o o	o o o	o o o o o o o o o o	o o o o o o o o o o	o o o o o o o o o	
5 years ago before protection	o o o o	o o o o o o o o	—	o o o o o o		o o	o o o	
Today after protection	o o o o o o	o o o o o o	—	o o o o o o o	—	o o o o o	o o o o o	
Projected need	o o o o o o o o o o o o	o o o o o o o o o o	o o o o o o o o o	o o o o o o o o o	o o o o o o o o o	o o o o o o o o o o	o o o o o o o o o o	
Ecological niche/ successional stage								?
Management steps	thinning, short rotation alternative sources	low spreading crowns or coppice shoots	longer rotation for seed bearing old trees	shrubs, root suckers pruning	long rotation high forest	high humus content no leaf sweeping	total protection enrichment planting?	

F Appendix: Discussion Questions

- If you were asked by the Minister of Forests to draw up guidelines for a new forest policy for your country what would be the key features of this policy. Make sure that you list what the objectives of your forest policy are to be. Provide an argument as to why forests should or should not be retained in areas where forest land has greater agricultural than forest potential.

- For your own country list out the major interest groups who would be affected by the forest policy you have just outlined. List the positive and negative effects on these interest groups, and indicate how you would change the policy to ensure that the negative consequences can be reduced.

Interest Group	Interest	Positive	Negative	Change in Policy

- 'The current community forestry policy is open, participatory, decentralised, and pragmatic' (A.L. Joshi, quoted in Talbott and Khadka, 1994).

Discuss this statement reflecting on the case-study material presented in Chapters 4 and 5?

- Does overpopulation necessarily lead to degradation of resources?

Discuss with reference to material presented in the Study Guide and from your own country experience.

Note: Key additional source references to provide background reading for this question: Fox, 1993; Tiffen et al. 1994

- Why do governments set a target figure for forest cover? Is it helpful? What are the implications for alternative land-use systems?
- Can Forest Departments respond to the demand for change being fostered at local level? Are financial and human resources available for wide-scale expansion and follow-up support of such approaches? Do government departments have adequate staffing or incentive structures to allow staff to function effectively at the local level?
- Will local forest institutions become unrepresentative and skewed against those who most need access to forest products?

Discuss the above statements in the light of the evidence provided in the Guide and drawing on your own experience.

The Four Key Actors in Forest Policy Development

Look at these four statements concerning the roles of the major actors in the forest sector. Discuss a) whether these are the major actors and the validity of each description; b) how to ensure that each provides a balanced input to the development of forest policy; and c) look at the relative role and influence of each of these groups in the formation of forest policy over the last 100 years.

Wildlife Conservationists[1]

An interest group with an influence on policy wholly out of proportion to their numbers. While their focus in practice has been on the preservation of unspoilt nature, defenders of the wilderness are prone to advance moral, scientific and philosophical arguments to advance their cause. Conservationists have found strong support from recent biological debates. The theme of biological diversity as an essential component of a direct and indirect, known and yet to be discovered survival value for humanity as well as an emphasis upon the 'intrinsic' rights of non-human species, has been prominent in recent debates on the preservation of wilderness areas. Wildlife conservationists share, with senior bureaucrats in particular, a similar educational and cultural background, and this proximity has in no small way influenced the designation and management of wildland. But for all their talk of the rights of 'non-human nature', most conservationists have been deeply insensitive to the rights of villagers displaced by national parks or whose access to forest produce has been curtailed by their constitution.

Timber Harvesters

Industrial exploiters of the forest resource stand at the opposite extreme to the wildlife conservationists. Industrial exploitation of the forest often leads to a substantial and even irreversible modification of the ecosystem. They are fashioning policy through the claim of an intimate link between industrialisation and prosperity. In the past state management of forests and heavy subsidies of raw materials ensured industrialists an abundant and cheap means of production. In an atmosphere of withdrawal of subsidies, forest industrialists are now agitating for the release of degraded lands for their exclusive use as 'captive plantations', with the added rationale that the use of these plantations will reduce the pressure on natural forests.

Rural Social Activists

These activists work to protect the rights of groups whose livelihoods are heavily dependent on forests. Some activists call for a radical reorientation of forest policy, so that it would more directly serve the interests of subsistence peasants, tribals, nomads and artisans. Others go further in asking for a total state withdrawal from forest areas; these can then revert to the control of village communities, which they believe have the wherewithal to manage these areas sustainably and without friction. There are many debates ongoing in India as to the role of the activists and indeed whether the claims they have made in the struggle for control over forest lands are justified.

Scientific Foresters

The brief of the Forest Department might be defined as the adjudication, in an ostensibly scientific and objective manner, of the competing claims of the above three interest groups . . . Foresters have often tended to act in narrow self-interest; holding on to control over forest land and the discretionary power that goes with it . . . (I)n practice the Indian Forest Department has consistently put industrial exploitation ahead of both ecological integrity and the rights of local communities.

Note

1. Derived from: Guha (1994).

G Appendix: Participatory Forest Management: Videos

Verity Smith

Introduction

This appendix has been included as a reference list for those wishing to design courses based around case-study material relevant to participatory forest management. The use of videos as a means of communication in both teaching and extension practice has been demonstrated to be of wide use.

The Use of Videos as a Medium of Communication in the Village

Each evening we went to another village. First there was an introduction and when it began to get dark the film was shown. Hundreds of people attended each session. Some documentary films did not hold people's attention. One film about the problems of forest protection, which was a well-recognised story about an illegal wood cutter who gave up his illegal activities, released the audience's emotions. The audience could be seen holding its breath, giving cries of joy and laughing. At the end of the showing there were so many requests from other villagers that we had to hold up to three sessions per evening.

The results of such programmes are hard to assess. However, it is certain that lower castes, women and children are reached by these films, since they are all free after the evening meal to participate and appeared to be involved in the message of the film.

Source: VSO volunteer reports

The Video Resource

Each entry provides a brief summary of the content, including a critical comment about its strengths and weaknesses, as an aid to helping foresters to develop their forestry extension skills. The geographical coverage has been extended beyond Nepal and India in order to cover some of the more generic issues. In all cases, contact names have been provided from whom to obtain further information about the availability of the videos.

Title:	Growing Links in Forestry
Country:	Pakistan (North West Frontier Province)
Language:	English
Length:	27 minutes

Description: This video explores ways in which the concept of people's participation is being developed in the forestry sector in NWFP, Pakistan. It shows how these processes are being implemented by 3 bilateral projects working with the NWFP Forest Department. The projects operate in diverse forest conditions, among different social groups with varying and complex land ownership and use patterns. They predict that, in situations of rapid resource degradation, the processes of social forestry being tried here provide the only way forward. Integrated local resource management systems are being developed through co-operation between departments, projects and local people.

Comments: This video was made to raise awareness amongst an 'educated' audience, and would be of interest to those who influence the development of forest policy. It is also relevant to all those involved in working at the field level and to those responsible for training. It illustrates briefly the different forms of participation during the project cycle: participatory village-level planning, implementation (including sustainable forest management, women's participation and community checkpoints for forest protection) and harvesting (using skylines). There are interesting comments about management plans being made by the relevant 'social unit', rather than being based solely on physical criteria. The importance of training in social aspects of forestry is clearly illustrated showing the need for new skills and active learning approaches.

Contact: Social Forestry Project
PO Box 9
Saidu Sharif
Pakistan
Tel: +92 936 711025
Fax: +92 936 720708

Title:	Farming with Trees
Countries:	Zambia, Uganda, Peru and Indonesia
Language:	English
Length:	18 minutes

Description: This video was produced in 1995 by ICRAF (International Centre for Research in Agroforestry). It questions the strict division of land between forest and farm, suggesting that a blurring of these frontiers leads to more sustainable land use and can contribute to food sustainability. Case studies are given of four farming families from tropical forest areas which are now suffering from ecological problems and poverty due in part to loss of trees. The video shows how they can profit through a variety of agroforestry technologies.

In S.E. Zambia, where loss of open woodland has led to soil degradation and poor harvests, unproductive land is fallowed with nitrogen-fixing *Sesbania*, increasing maize yields and providing local firewood on farm land. In S. Uganda, people have had to resort to farming on steep hillsides leading to loss of soil. Agroforestry experiments with contour planting of *Calliandra* can be successful, if farmers are prepared to work collectively.

The Peruvian example highlights both the environmental and agricultural hazards of slash and burn farming as practised by migrant farmers. In the example from Sumatra, in Indonesia, intricate and productive slash and burn systems are presented. Trees in the forest gardens, developed under these systems, are valued for non-wood forest products such as medicine, resin and fruit.

The combination of farmers' traditional wisdom and modern science is advocated as a way to make farming more effective. The video emphasises the need to convince both policy-makers and farmers that working with trees is a better option than felling them.

Comments: A useful video to show foresters, agriculturalists and policy-makers. The use of case-study families is convincing and agroforestry technologies are shown as both helping to reduce poverty now and assisting towards sustainable land use for the future. Various social issues are raised, both as causes and results of tree loss and could be used to stimulate group discussion. There are good examples of indigenous technical knowledge. In addition to this programme, the cassette follows on with a more in-depth analysis of some of the farming problems considered in the first part.

Contact: Television Trust for the Environment
Distribution and Training Centre
Postbus 7
3700 AA Zeist
The Netherlands
Tel: +31 3404 20499
Fax: + 31 3404 22484

Title:	A Participant's Diary of a PRA Exercise
Country:	India, Karnataka
Language:	English
Length:	20 minutes

Description: This video shows the use of participatory rural appraisal (PRA) on a watershed management project. The PRA tools illustrated include:

- do-it-yourself 'warm up' exercise as an 'equaliser'
- seasonality diagramming of rainfall and labour
- mapping (of watershed)
- model
- transect walk (observation and informal interviews)
- matrix ranking (of local tree varieties and uses)
- time line chart
- oral history techniques (interviewing an older person about village history)
- Venn/chapatti diagrams
- relationships between local institutions
- seeking information on non-farming livelihoods
- wealth ranking

The whole exercise, which took several days, is described as being informal and open, aiming at a joint approach to decision-making and problem solving.

Comments: This is a very brief run through several PRA techniques. It might have provided a more useful introduction to PRA techniques if more time had been spent on a smaller number. However, it does introduce the terms and the initial 'warm-up' exercise is a good idea which could be copied in training or a real village situation.

Contact: MYRADA
2, Service Road
Dolmur Layout
Bangalore 560 071
Karnataka
India
Tel: +91 812 567166

Title:	Forestry and Food Security
Countries:	Ghana, Amazonia and Thailand
Language:	English, French, Spanish
Length:	14 minutes

Description: This FAO video uses case studies from 3 countries. It emphasises how many foods come directly from trees and forests, providing both a direct and indirect source of income and energy. The nutritional role of forests is often underemphasised. Trees provide nutritional supplements, basic survival needs in the 'hungry season' and also medicines. It touches briefly on the need for confidence in tree tenure in order to encourage people to plant trees. It gives examples of the value of agroforestry and encourages wide-ranging links between the forestry, agricultural, nutrition and livestock sectors. The video ends by stressing the need to consider nutrition issues in project designs, management plans and national development policies so that the contribution of forestry to food security can be strengthened.

Comments: The video is composed of slides enhanced by computer graphics. The language is simple and persuasive. As a training tool it could be used at a more basic level with schoolchildren to promote discussion of the use of food from trees It could alternatively be used to introduce a deeper discussion on land and tree tenure issues, agroforestry links, survival foods, small income-generation projects from trees and biodiversity.

Contact: For further information contact the Forestry Department at FAO or the regional focal point for the Forests Trees and People Programme (addresses given at the end of this appendix).

Title:	Gender Analysis for Forest Development Planning
Country:	various, see below
Language:	English, French (forthcoming)
Length:	19 minutes

Description: This FAO video makes a strong case for 'gender analysis' in forest development planning. It illustrates how gender roles may vary from area to area, emphasising the need for careful identification of the needs and priorities of men and women whilst developing new programmes. Eight case studies are presented from six countries, each with a different natural resource perspective:

Nepal	Watershed management
Bhutan (2)	Integrated land-use management
India (2)	Non-wood forest products and fuelwood
Thailand	Reforestation of denuded land
Bangladesh	Agroforestry and nursery programmes
Sri Lanka	Income generation and tree-related systems to improve the quality of life

The video presents four key questions to be addressed in a

gender analysis:

1. What is getting better and what is getting worse? the 'development profile'
2. Who does what? the 'activity profile'
3. Who has what? the 'resources profile
4. What should be done? the 'programme action profile'

Comments: This video uses slides enhanced by computer graphics. It is a useful video emphasising joint planning between villagers, foresters and planners. It cautions against the use of stereotypes which assume certain gender roles. This video can be used to question assumptions more generally, and provides a useful background to discussions about the relative roles of women and men in natural resource management.

Contact: For further information contact the Forestry Department at FAO or the regional focal point for the Forests Trees and People Programme (addresses given at the end of this appendix).

Title:	Handing Over The Stick
Country:	Tanzania
Language:	English
Length:	31 minutes

Description: This video illustrates the potential of local (people's own) institutions, and the importance of customary rights and empowerment of local people for sustainable natural resources management. It stresses the value of local knowledge, both of natural resources and of social institutions. It describes villagers as the expert researchers to whom professionals need to listen and then share their own information to resolve environmental problems. The video shows some participatory rural appraisal (PRA) techniques being used in a school and in villages, including participatory mapping, a transect walk and village history. The PRA exercise resulted in the revitalisation of the Dagashida – an endogenous village-level institution which resembles a village parliament. Two such meetings are shown where villagers discuss environmental degradation and how they might address these problems.

Comments: This video has relevance to a wide range of people. The emphasis on local knowledge and social institutions of decision-making and control would also interest rural development workers and anthropologists. The commentary of the video speaks directly to village people, thanking them for sharing their traditional knowledge and recognising its value for the younger generation and for professionals. This video could be used to raise issues to be followed up in group discussions.

Contact: The video was made by Scandinature Films in cooperation with FAO, Forests, Trees and People Programme (FTPP), and

Swedish International Development Authority (SIDA).

For further information contact the Forestry Department at FAO or the regional focal point for the Forests Trees and People Programme (addresses given at the end of this appendix).

Title:	What is a Tree?
Country:	several African countries
Language:	English
Length:	17 minutes

Description: This FAO video, made from slides enhanced by computer graphics, starts with an African village storyteller explaining the cultural and traditional importance of trees to rural people. A tree is explained as being many things at different times to different people. Foresters need to recognise all the values of a tree to rural people. Therefore they need to ask appropriate questions to find out how trees are used and then combine their own professional knowledge with the traditional knowledge and skills of rural people. The video stresses the importance of non-timber forest products as well as timber. Uses mentioned include: cultural (magic, ritual and ceremonial), food, building, medicinal, transport, soil conservation and income generation. It alludes to gender differences but does not emphasise gender analysis. It considers briefly the different qualities of wood which influences its utility, for example non-porous wood for boat building and termite-resistant wood for house building.

Comments: This is a useful, simple introduction to NTFPs. The video is unusual in mentioning the cultural importance of trees in the social life of rural communities. This video could be used by trainees to list farmers' criteria for choosing which trees to plant.

Contact: For further information contact the Forestry Department of FAO or the regional focal point for the Forests, Trees and People Programme (address given at the end of this appendix).

Title:	Gaunle Ko Ban Byavastha
Country:	Nepal
Language:	English and Nepali versions
Length:	7 minutes

Description: This video was made in 1990 by NACFP for training Forest User Groups. It counters the belief that all the forests are disappearing. Many trees are grown on private land and some patches of forest are managed by people co-operatively. User groups already exist in many places and these should be recognised and built upon. The video examines three local systems of management in different villages in Central Nepal. Commenting on these, it is clear that systems may change over time and that neither the committee nor a watcher is essential. The two fundamentally important elements are:

1. that there is a defined group of users
2. that people agree on what they may do and use

Field staff are advised, firstly, to identify any existing user group and then to make sure there is adequate discussion within the whole user group before any committee is formed. There is ample evidence that local people can manage their forests.

Comments: Seeing this video could be quite a confidence-building experience for local people whose indigenous management systems are recognised as beneficial to the forest. It also serves to remind field staff of local capabilities and structures that already exist. Well understood agreements between users are more significant than committees which may exist only in name or do little to protect the forest. Fieldworkers seeing this can be encouraged to contribute examples of local management systems they have come across in their own countries.

Contact: ANUTECH Pty Ltd
Forestry and Environment Division
Canberra ACT 0200
Australia
Fax: +616 249 5875
Tel: +616 249 5672

Title:	Naya Bihani
Country:	Nepal
Language:	Nepali
Length:	1 hour

Description: This video shows a young Forest Ranger integrating into a village and building rapport with the local people. He demonstrates the skills of listening, observing, enabling the village people, including the women, to participate in decisions about trees and rural development.

Comments: It is made in the style of an entertainment, with a strong story line, somewhat like an Indian movie. Two issues are central: tree management and remarriage of widows. Since the language of the video is Nepali, it may have limited use for non-Nepali speakers, although the message is well conveyed through the visual image.

Contact: Community and Private Forestry Division
Hatisar
Kathmandu
Nepal
Tel: +977 1 411594/414004

Title:	Community Forestry – For Whose Benefit?
Country:	Nepal
Language:	English & Nepali versions
Length:	41 minutes

Description: This video is designed for training Forest User Groups and was made in 1993 for the NACFP. It shows the need for careful management of the forest in order to meet both present and future needs for forest products. It explains the new role of Forest Department Rangers: to work with rural people in helping them manage their own forest, giving technical advice, assisting with Operational Plans and helping villagers consider the needs of *all* members of the Forest User Group. The Forest Department staff can also help with negotiations and resolving disputes if necessary. The stages of creating an Operational Plan and the advantages of working together as a community are clearly spelt out. These may include access to more forest products, a saving of time collecting them and the possibility of accumulating funds for locally agreed development projects.

Comments: This presents a convincing case for the management of forests by Forest User Groups. It is targeted at villagers but could also be used with Forest Rangers to help them consider the knowledge, skills and attitudes demanded of them in their new role. The need for communication and problem-solving skills, as well as technical ones, is evident. Much of the commentary is direct speech by a variety of villagers, men and women, which adds to the video's credibility.

Contact: ANUTECH Pty Ltd (address given above)

Title:	A Tool Kit For Community Forestry
Country:	Nepal
Language:	English and Nepali versions
Length:	32 minutes

Description: This video starts from an explanation of the new roles of Forest Rangers – to facilitate new forest management partnerships between forest users and Forest Department staff. Forest Rangers must be able to supply information and advice to help people to help themselves and also be willing to *learn* from rural people. Fieldworkers need information about tree-related resources and the needs of the various people who live locally. This information-gathering process must be simple, efficient, accurate and reliable; a case is developed for the use of Participatory Rural Appraisal (PRA) skills.

The following 'toolkit' of communication and information-gathering skills is contrasted with the traditional forester's toolkit of physical measuring instruments. Each of the 8 PRA skills described is seen in practice and discussed.

1. Establishing rapport
2. Informal interviews
3. Reaching women
4. Key informants
5. Participatory mapping
6. Participatory forest profile
7. Time chart
8. Direct observation

Comments: This video is presented in very clear, simple language. The video is not exclusively about the information-gathering techniques which are now associated with PRA. The basic communication and relationship-building skills are also emphasised. It touches on attitudes and skills and is likely to build confidence in staff who are asked to work in a new participative way. It is particularly relevant, even essential, viewing for forestry staff at field level. However, any level of Forest Department or NGO staff would find it interesting. Although aimed at the forestry sector, it may well also be appreciated by other professionals working in agricultural extension or rural development. This is a professional production highly recommended by viewers from various Asian and African countries.

Contact: ANUTECH Pty Ltd (address given above)

Title:	Community Forestry – Village's Wealth
Country:	Nepal
Language:	English and Nepali versions
Length:	31 minutes

Description: This video contains 4 programmes and was made in 1993 by the NACFP for Nepal Television. Programmes 2, 3 and 4 contain clips of material from the longer NACFP video 'Community Forestry – For Whose Benefit?'

Programme 1. Improved Stove Programme (8.58 minutes)

This extension video highlights the disadvantages of the traditional village fireplace as unhealthy and inefficient, contributing to the pressure on the forests and adding to women's already high workload through the need for constant wood collection. It explains that the new improved stoves can help to lessen these problems and also improve the flavour of food. It shows briefly the technology of making these stoves from locally available materials. The commentary stresses the advantages of the stoves: they cost less, can be constructed to the optimal dimensions for each family and can be easily repaired when necessary.

This video could be used with staff at forest ranger level to raise awareness of the different issues which should be considered when trying to change behaviour (for example, using a new type of stove), knowledge, understanding, skill and attitude.

Programme 2. Women in Community Forestry (6.44 minutes)
This video emphasises women's dependence on the forest for domestic and livestock uses. Any changes in forest use, by means of altering the Operational Plan, should not increase their workload. Women have a deep knowledge of forests and forest products so they should be encouraged to contribute this in local decision-making. This will help to produce benefits for the present and to conserve healthy forests for the future.

Training use of this video could include discussion of women's many forest-related tasks and their specific forest knowledge and skills. If groups of Forest Rangers are asked to make up the time-table of a typical day of a rural woman there are likely to be some interesting discussions and a few surprises. This is particularly the case if a group of field staff from different countries are participating on the same training course, where cultural roles and perceptions may be very different.

Programme 3. Why Community Forestry? (7.20 minutes)
The proverb 'Green forests are Nepal's wealth' is a reminder of people's dependence on forest products. The Nepalese Government's Community Forestry Programme aims to help people to obtain the products they need, without damaging the forest. The new forestry legislation permits forests to be handed over to local Forest User Groups. By developing management plans they are involved in the protection, conservation and management of their forest. They must also ensure that benefits are distributed fairly. These benefits may later include funds from the sale of forest products, which can go towards locally agreed development projects. The objective is collaboration between Forest User Groups and the Government.

This video is a basic introduction to the Community Forestry Programme and is of general interest.

Programme 4. The Role of the Forest Department (8.50 minutes)
Under the Community Forestry programme, Forest User Groups can be given legal authority to manage their forest. They need to make an agreement with the Forest Department, known as the Operational Plan, in which the general management decisions and details of the extraction of benefits are stipulated. The video goes on to describe the new role of the Forest Rangers. They can give technical advice on pruning and harvesting, helping with the development of the Operational Plan to ensure that women's workload is not increased through the Plan. The Ranger assists members to negotiate amongst themselves so that they all get a fair share of the forest benefits. A central aim of the retraining of Forest Department staff is to increase co-operation between villagers and the Department.

This video follows on well from the previous Programme 3. It provides Forest Rangers with a good introduction to the issue of different stakeholders or interest groups at village level and the need to develop negotiation skills to help them resolve disputes over benefits.

Contact: ANUTECH Pty Ltd (address given above)

Title: Participatory Research with Women Farmers
Country: India, Andhra Pradesh
Language: English
Length: 22 minutes

Description: This video illustrates the key role women can play in extending and conserving genetic diversity in farming communities. It shows joint research being done by scientists at the gene bank at the International Crops Research Institute for Semi-Arid Tropics (ICRISAT), and women farmers on pigeon peas. The approach is decentralised and participatory. On-farm trials are followed by evaluations by the farmers jointly with scientists, facilitated by a social scientist. Various participatory rural appraisal (PRA) techniques are shown: farm walks and informal interviews plus two ranking methods: pair-wise ranking and direct matrix ranking.

Comments: This video has various uses in training. It encourages discussion on research taking take place on farmers' fields, supported by scientists. It shows the value of local agricultural knowledge, which may well mean involving women in the research. The second half of the video is useful in clearly illustrating several PRA techniques and promoting a participatory approach at village level.

Contact: Television Trust for the Environment
Distribution and Training Centre
Postbus 7
3700 AA Zeist
The Netherlands
Fax +31 3404 22484
Tel +31 3404 20499

Title: Sadupayog II Forest Management for People
Country: Nepal
Language: English and Nepali versions
Length: 17.17 minutes

Description: This video was made for NACFP in 1989 and then updated in 1995. It promotes the wise use of forest resources under the Nepal Government's community forestry legislation, under which, many forests are being transferred to management by local users. It takes the viewer through the stages of this transfer, emphasising the processes which must take place between the villagers and the Forest Ranger. These are:

1. Gathering information
2. Negotiations
3. Writing the Plan

It explains that Forests Rangers need new skills for this work: the ability to give technical advice on ways to promote locally desired forest products and the communication skills of interacting with villagers.

Comments: Obviously a useful training video for Forest Rangers in Nepal. It could, however, also be helpful on a training course in other countries when examining participative or 'bottom-up' planning in contrast to the more common 'top down' methods. It emphasises the new relationships and attitudes required when seeking information from forest users and the importance of careful negotiation. It would be a good video to precede the NACFP video 'A Toolkit for Community Forestry' which looks at information gathering skills in more detail.

Contact: ANUTECH Pty Ltd (address given above)

Title:	Ban Sambardhan Silviculture for User Groups
Country:	Nepal
Language:	English and Nepali versions
Length:	17 minutes

Description: This video was made in 1990 for the NACFP to train field staff in silvicultural treatments which can be carried out by villagers themselves. Two basic ideas underlie the treatments:

1. The forest is used but not destroyed
2. The forest structure and species composition can be modified for the future

Forest Department field staff must first find out what products the various types of users want and then advise on appropriate silvicultural treatments to obtain them. These simple treatments need to be agreed with the users: what to cut, what to leave, what to treat. The video then gives fairly detailed silvicultural advice for pine plantations and also shrubland. The pruning instructions are aided by clear graphics.

Comments: This is interesting since it is mainly a technical video, but it adapts the forestry advice to community needs and preferences. It also recognises the importance to users of non-timber forest products, including grasses. It reminds foresters that villagers are knowledgeable and well able to carry out these treatments, indeed they may be doing so already. It is a useful video including discussion about the management of natural forests, complementing the more frequent teaching on nurseries and plantations.

Contact: ANUTECH Pty Ltd (address given above)

Title:	Sustainable Community Forestry
Country:	Nepal
Language:	English
Length:	30 minutes

Description: This video is about self-governance in action. It presents the story of a village Forest User Group managing a degraded forest. During the previous three years, under the care of the villagers, the forest has gradually regenerated sufficiently to permit periodic harvesting of fuel wood and fodder. This success has encouraged the villagers to develop clean drinking water systems, a representative management system which includes women and an income-generating project.

Comments: The video is about the Nepal Resource Management Project managed by United Mission to Nepal (UMN). It is a useful case study to analyse through group work to discuss what the strengths and weaknesses are of the approach adopted by the project.

Contact:
International Center for Self-Governance
Institute for Contemporary Studies
720 Market Street
San Francisco
California 94102
USA
Fax: +415 986 4878
Tel: +415 981 5353 Ext. 226

For copies of the FAO videos please contact the nearest regional focal/distribution point:

Asia
FTPP at RECOFTC
Regional Community Forestry Training Centre
c/o Faculty of Forestry
Kasetsart University
Bangkok 10903
THAILAND

Europe (English)
The Editor FTPP Newsletter
Swedish University of Agricultural Sciences
Box 7005
75007 Uppsala
SWEDEN

East Africa
FTPP Network Coordinator
Forest Action Network
PO Box 21428
Nairobi
KENYA

Francophone Africa
FTPP Regional Facilitator for Francophone Africa
Institut Panafricain pour le Developpement
IPD-AC
BP 4078
Douala
Cameroon

Latin America and Caribbean (Spanish)
Revista Bosques, Arboles y Comunidades Rurales
c/o Carlos Herz – Ediciones ABYA-YALA
12 de Octubre 1430 y Wilson
Casilla 8513
Quito
ECUADOR

North America and Caribbean (English)
International Society of Tropical Forests
5400 Grosvenor Lane
Bethesda, Maryland 20814
USA

Other regions
The Senior Community Forestry Officer
Forestry Policy and Planning Division
Forestry Department
Food and Agriculture Organisation of the United Nations
Viale delle Terme di Caracalla
Rome 00100
ITALY

H Appendix: Information Sources[1]

Newsletters and other information

This is not an exhaustive list of newsletters available but provides information about those that regularly contain articles on participatory forestry or related issues.

The Asia Forest Network

This network was formed in 1991 to link field researchers and policy-makers, NGOs and donor agencies, committed to community involvement with forest regeneration and protection. Supported by the Berkeley-based Secretariat and regional offices in Manila and New Delhi, network members throughout Asia are actively engaged in documenting traditional and emerging community forest management practices. Network members are developing new methodological tools and through the network communicating strategies and successes.

For more information contact:
Mark Poffenberger
Center for Southeast Asia Studies
University of California, Berkeley
2223 Fulton, Rm. 617
Berkeley, California 94720
Tel: 510 642 3609
Fax: 510 643 7062

Biodiversity Conservation Network

Biodiversity Support Program
Biodiversity Conservation Network
151-B Gonzales Street
Loyola Heights
Quezon City
Philippines

The Common Property Resource Digest
Published by the Department of Agricultural and Applied Economics, University of Minnesota, 332e C.O.B. 1994 Buford Avenue, Saint Paul, Minnesota, 55108, USA

Forest Conservation Programme Newsletter
Published by IUCN
Available from:
IUCN Forest Conservation Programme
rue Mauverney 28
1196 Gland
SWITZERLAND

Forests, Trees and People Newsletter
Published by International Rural Development Center of the Swedish University of Agricultural Sciences, Uppsala and the Community Forestry Unit of FAO; copies are sent free of charge upon request to IRDC/SUAS Box 7005 S-750 07, Uppsala, Sweden.

Haramata
Quarterly newsletter of the Dryland Programme of the International Institute for the Environment and Development
Available from: IIED, 3 Endsleigh Street, London WC1H ODD, UK

ICIMOD International Centre for Integrated Mountain Development
Useful publications on a range of natural resource management issues in the montane areas of the world.
For further information on publications write to:
The Publications Unit
ICIMOD
GPO Box 3226
Kathmandu
Nepal

IDRC Notes
Informative articles on development issues.
Published by the International Development Research Center of Canada; copies are sent free of charge on request to Communication Division, IDRC PO Box 8500, Ottawa, Canada K1G 3H9

ILEIA Newsletter
The newsletter of the Information Centre for Low External Input and Sustainable Agriculture, which disseminates research findings and field experiences, organises workshops and supports regional networking activities. Publication available from: ETC Foundation, Kastanijelaan 5, PO Box 64, 3830 AB Leusden, The Netherlands

Rural Development Forestry Network

Newsletter and thematic papers published by the Rural Development Forestry Network, Overseas Development Institute, Regent's College, Inner Circle, Regent's Park, London NW1 4NS, UK

Wasteland News

Useful newsletter about new approaches in participatory forest management in India.
Available (for small cost) from:
Society for Promotion of Wastelands Development
Shriram Bharatiya Kala Kendra Building
1 Copernicus Marg
New Delhi 110 001
INDIA
Also available are a small number of very interesting, well-written, working papers on issues in participatory forest management (write to above address for further information)

Note

1. Some of the information sources in this Appendix are reproduced from Borrini, 1992

Index